INTERNATIONAL CENTRE FOR MECHANICAL SCIENCES

COURSES AND LECTURES - No. 48

ANTONI K. OPPENHEIM

UNIVERSITY OF CALIFORNIA

INTRODUCTION TO GASDYNAMICS
OF EXPLOSIONS

COURSE HELD AT THE DEPARTMENT
OF HYDRO- AND GAS-DYNAMICS
SEPTEMBER 1970

UDINE 1970

SPRINGER-VERLAG WIEN GMBH

ISBN 978-3-211-81083-5   ISBN 978-3-7091-4364-3 (eBook)
DOI 10.1007/978-3-7091-4364-3

PREFACE

Over the last half a century, Fluid Mechanics, in its evolution from Aerodynamics, was encompassing within its scope more and more physical phenomena. The Boundary Layer Theory brought under consideration the effects of viscosity and heat conductivity ; Gasdynamics took into account the effects of compressibility ; the Aerothermochemistry included the influence of chemical composition of the medium, bringing into consideration the concomitant processes of diffusion and chemical reaction : Gasdynamics of Rarefied Media embodied the consequences of the particulate nature of matter ; and the Magnetohydrodynamics concerned itself with the effects of electromagnetic forces and the associated with them energy transformations.

Starting from the treatment of compressible boundary layers, the theory of all these subjects had to be concerned with some aspects of energy transfer . In most cases, however, the kinetic rates at which energy conversions occur in nature did not play a dominant role. The natural next step in the evolution of Fluid Mechanics appears therefore to be the consideration of flow phenomena that take place when the rate of energy conversion becomes significant. In nature this is the case in explosions where, as a rule, the energy conversion system is exothermic

while the flow field is non-steady. This is then the
domain of Gasdynamics of Explosions a relatively new
branch of Fluid Mechanics that is concerned primarily
with the establishment of the interrelationship between the rate processes of energy deposition at a
high-powered density level in compressible media and
their concurrent non-steady motion.

Presented here first are the phenome-
nological aspects of the subject based mainly on high
speed schlieren records of wave processes in detonative gas mixtures. This is then followed by the dis-
cussion of the fundamental principles and methods used
for the interpretation of these phenomena. Of particular importance in this connection is the analysis of
gasdynamic discontinuities and their interactions,
and the blast wave theory in its more general sense
than just the conventional case of decaying explosions.

The scientific significance of this
subject stems from the fact that it provides us with
a unique facility for the study of direct macroscopic
effects of exothermic processes occurring at high
rates. As to physical systems for which such knowledge is of particular importance, one has the whole
spectrum of propulsion technology, from internal combustion engines to rocket boosters, as well as some
basic aspects of energy conversion, again covering
an extremely wide range of applications from spark
ignitions to cosmic explosions.

The exposition of the subject matter

in this book has been based primarily on results of a research program which was conducted for over a decade at the University of California in Berkeley. The success of the studies carried out in this connection is thus due in great measure to the contributions of the many students and collaborators with whom the author had the good fortune of being associated during that time. It is with great pleasure, therefore, that he would like to take this opportunity in order to express his thanks for the valuable help he received from Drs. G.J. Hecht. R.A. Stern, A.J. Laderman, P.A. Urtiew, J.R. Bowen, R.W. Getzinger, C.W. Busch, R.L. Panton, W.N. Podney, L.J. Zajac, J.J. Smolen, E.A. Lundstrom and Messrs. M. M. Kamel, A.L. Kuhl, W.J. Meyer and K. Hom who contributed directly to the contents of this work ; to Miss K. Tomlinson who exhibited great patience and skill in typing the many versions of the manuscript, and to his wife and daughter who put up with him graciously at home when he was working on the text.

The author wishes to acknowledge also his appreciation for the support he received from the Air Force Office of Scientific Research under Grant AFOSR 129-67, the National Aeronautics and Space Administration under Grant NsG-702/05-003-050 and the National Science Foundation under Grant NSF-2156.

A.K. Oppenheim

Contents.

Contents

Nomenclature.

Dimensional Quantities.

a - local velocity of sound ; conservation quantity in Eq. (4.1)

A - flow cross-section area

b - conservation quantity in Eq. (4.1) ; segment in Fig. (3.3b)

$\dot{c}$ - rate of change of conservation quantity in Eq. (4.1)

e - specific internal energy

f - stream force per unit area

F - force

h - specific enthalpy

$\dot{m}$ - flow rate per unit area

M - mass

p - pressure

P - power

r - space coordinate

s - entropy

t - time

T - temperature

u - flow velocity

w - wave velocity

v - relative wave velocity

V - velocity

α - triple-point trajectory angle

δ - flow deflection angle

ϑ - incidence angle

ϱ - density

Φ - angle between two fronts in a triple-wave inter-
     section defined by Fig. (3.18)

Ω - source term introduced in Eq. (4.2)

Dimensionless Quantities.

A - velocity of sound ratio

A_k - conservation quantity defined by Eq. $(4.6)_1$

B_k - conservation quantity defined by Eq. $(4.6)_2$

c - exponents defined by Eq. (4.155)

C - parameter of the Hugoniot relation defined by
     Eq. (3.26)

C_k - conservation quantity defined by Eq. $(4.6)_3$

D - function defined by Eq. $(4.47)_1$

$\tilde{D}$ - function defined by Eq. $(4.69)_1$

$\hat{D}$ - function defined by Eq. $(4.94)_1$

f $=$ $\dfrac{u}{w_n}$

F $=$ $\dfrac{\tau}{x}\,f$

$\underline{F}$ - functions defined by Eqs. $(4.47)_2$, $(4.48)_1$ and
     $(4.49)_1$

$\tilde{\underline{F}}$ — functions defined by Eqs. $(4.69)_2$, $(4.70)_1$ and $(4.71)_1$

$\hat{\underline{F}}$ — functions defined by Eqs. $(4.94)_2$, $(4.95)_1$ and $(4.96)_1$

$g = \dfrac{p}{\rho_a w_n^2}$

$\underline{G}$ — functions defined by Eqs. $(4.47)_4$, $(4.48)_3$ and $(4.49)_3$

$\tilde{\underline{G}}$ — functions defined by Eqs. $(4.69)_4$, $(4.70)_3$ and $(4.71)_3$

$\hat{\underline{G}}$ — functions defined by Eqs. $(4.94)_4$, $(4.95)_3$ and $(4.96)_3$

$h = \rho/\rho_a$

$H = \dfrac{h}{p_j/\rho_j}$

$\underline{H}$ — functions defined by Eqs. $(4.47)_3$, $(4.48)_2$ and $(4.49)_2$

$\tilde{\underline{H}}$ — functions defined by Eqs. $(4.69)_3$, $(4.70)_2$ and $(4.71)_2$

$\hat{\underline{H}}$ — functions defined by Eqs. $(4.94)_3$, $(4.95)_2$ and $(4.96)_2$

$j \equiv \dfrac{\partial \ln A}{\partial \ln r}$ — geometric index (0 for plane flow, 1 for cylindrical, 2 for spherical)

$\mathcal{J}$ — the Jacobian operator

k — index (0 for conservation of mass, 1 of momentum, 2 of energy)

K — conservation quantity in Eq. (4.16)

M — Mach number ; conservation quantity in Eq. (4.16)

N — conservation quantity in Eq. (4.16)

$P \equiv \dfrac{p_\ell}{p_i} = \dfrac{p_n}{p_a}$

q — energy deposited per unit mass

$Q \equiv \dfrac{g}{p_i/\rho_i}$

$U \equiv \Delta u / a_i^2$

$x \equiv r / r_n$

$y \equiv a_a^2 / w_n^2$

$Z \equiv \left(\dfrac{\tau}{x}\right)^2 \dfrac{g}{n}$

$\underline{Z}$ — functions defined by Eqs. $(4.47)_5$, $(4.48)_4$ and

    $(4.49)_4$

$\underline{\tilde{Z}}$ — functions defined by Eqs. $(4.69)_5$, $(4.70)_4$ and

    $(4.71)_4$

$\underline{\hat{Z}}$ — functions defined by Eqs. $(4.94)_5$, $(4.95)_4$ and

    $(4.96)_4$

$\alpha \equiv \dfrac{2}{\rho}\left(\dfrac{\partial p}{\partial a^2}\right)_s = 2\left(\dfrac{\partial h}{\partial a^2}\right)_s$; constant defined by Eq. (4.129)

β — asymptote of the Hugoniot hyperbola

γ — specific heat ratio

$\Gamma \equiv \left(\dfrac{\partial \ln p}{\partial \ln \rho}\right)_s = \left(\dfrac{\partial \ln h}{\partial \ln e}\right)_s = \dfrac{a^2}{p/\rho}$

$\zeta \equiv \left(\dfrac{\partial \ln x}{\partial \ln \xi} + 1\right) - \dfrac{\tau}{x} f$

$\eta \equiv \dfrac{t_n}{t_a}$

$\varkappa \equiv -\dfrac{\rho}{p}\left(\dfrac{\partial e}{\partial \ln \rho}\right)$

$$\lambda \equiv -2 \frac{d \ln w_n}{d \ln \xi} = -2 \frac{d \ln a_a}{d \ln \xi} + \frac{d \ln y}{d \ln \xi}$$

$$\mu \equiv \frac{d \ln \xi}{d \ln \eta}$$

$$\mu' \equiv \frac{d \ln \mu}{d \ln \xi}$$

$$\nu \equiv \frac{\rho_i}{\rho_j} = \frac{\rho_a}{\rho_n} \quad ; \quad \text{constant defined by Eq. (4.127)}$$

$$\xi \equiv \frac{r_n}{r_a}$$

π – particle path

$$\rho'_o \equiv \frac{d \ln \rho_a}{d \ln \xi}$$

$$\sigma \equiv \frac{e}{w_n^2}$$

$$\tau \equiv \frac{t}{t_n}$$

$$\phi \equiv -\left(\frac{\partial P}{\partial y}\right)_n \quad \text{where } n = H, R, S \; ; \; \text{"field" coordinate}$$
$$(= x \text{ or } \tau \text{ or } f)$$

$$\varphi \equiv \frac{x}{\tau^\mu}$$

$$\Phi_E \equiv \frac{\Gamma}{\varkappa + 1} \frac{h}{g} \left[\omega_E - f \omega_F - \left(\sigma + \frac{g}{n} - \frac{f^2}{2} \right) \omega_M \right]$$

$$\Phi_F \equiv \frac{1}{f} \left[\omega_F - f \omega_M \right]$$

$$\Phi_M \equiv \omega_M$$

$$\tilde{\Phi}_E \equiv \frac{\Phi_E}{f}$$

$$\tilde{\Phi}_F \equiv \frac{\Phi_F}{f}$$

$$\tilde{\Phi}_M \equiv \frac{\Phi_M}{f}$$

$$\hat{\Phi}_E \equiv \Phi_E$$

$$\hat{\Phi}_F \equiv \frac{\Psi}{x^\delta g}(\omega_F - f\omega_M)$$

$$\hat{\Phi}_M \equiv \frac{\Psi}{f}\omega_M$$

$$\psi \quad - \quad \text{"front"coordinate} \; (= \xi \text{ or } \eta \text{ or } \gamma \;)$$

$$\psi_i' \equiv \frac{d \ln \psi}{d \ln \xi}$$

$$\Psi \equiv h x^\delta \zeta = \exp\left[\mu \int_n^\tau \omega_M \, d\tau\right]$$

$$\omega_i \equiv \Omega_i \frac{r_n}{w_n^{k+1}}$$

Subscripts.

o – denotes a reference point of the front trajecto ry

a – denotes the state of the ambient medium into which the wave propagates

E – refers to energy

F – denotes specific volume or velocity ratio asso-

ciated with constant pressure deflagration ;
refers to force

G —   denotes pressure ratio corresponding to combustion at constant volume

i —   denotes conditions immediately ahead of a gasdynamic discontinuity ; stands for M, F or E (in ω_i and Ω

j —   denotes conditions immediately behind a gasdynamic discontinuity

J —   refers to Chapman-Jouguet detonation

k —   stands for 0, 1 and 2 referring, respectively, to the conservation of mass, momentum and energy

K —   refers to the Chapman-Jouguet deflagration

P —   refers to constant pressure

M —   refers to mass

n —   refers to normal ; denotes the point of the front trajectory or the state immediately behind the front

N —   refers to the Von Neuman spike

v —   refers to constant specific volume

I —   denotes the weaker front in a triple wave intersection

II —   denotes the stronger front in a triple wave intersection.

Chapter 1.
Genesis and Sustenance.

The Birth and Life of an Explosion.

1. 1. Introduction.

As amazing as it may seem, the study of the birth and life of explosions, that is of the chemical and gasdynamic processes that are associated with the generation, growth and sustenance of explosions, escaped up to now, by far and wide, the attention of the scientific community.

This is particularly surprising since, on one hand, explosion processes are *de facto* occurring in most combustion systems, especially those used as energy sources for propulsion and, on the other, an enormous amount of effort that has been spent by manking on the development of explosives, both chemical and nuclear, for use as destructive weapons. Thus , in recent years, a lot of effort was spent on the acquisition of knowledge on the destructive effects of explosion waves or, as they are now referred to in technical literature, blast waves, that is phenomena which are associated essen-

tially with the period of decay of the process of explosion, while the study of their development, or
birth and life,that is phenomena associated with the
generation and sustenance of these processes, received
much less attention.

By means of modern scientific instru-
ments that exploit extremely short pulses of electrons
and photons- the latter, obtained by means of lasers,
ranging in length of light beam from a few feet down
to just a fraction of an inch-one is now in a position
to perform experiments where the actual birth and life
of explosion processes can be observed with a sufficiently good resolution, both in time and in space, to
reveal a significant amount of information about their
actual mechanism.

In this respect the present chapter ac-
complishes the following : (1) it describes how basical-
ly such phenomena are studied-a subject that became
recently known as the Gasdynamics of Explosions, (2)
it illustrates some of their more interesting technological applications, and (3) it outlines their exci-
ting future prospects of attaining a significantly
better control over the explosion process than it has
ever been deemed possible.

1. 2. Mechanics of Explosions.

The best understood form of the explo-
sion process is that associated with the bursting of
a wall of a vessel containing a high pressure gas. There
is, at first, formed a pressure wave that produces a
shock front moving into the surroundings, while,at the
same time, a rarefaction wave is generated propagating
into the high pressure source. The simplest form of
such a process is obtained by the use of a shock tube-
a device that gained over the recent years great popularity as means for experimental investigations of
chemical, thermodynamic ans gasdynamic phenomena that
occur in gaseous media at high temperatures-the key
tool, in fact, that paved the way for the development
of re-entry vehicles and is now ushering in the era
of hypersonic flight for the airplanes of the future.

The salient gasdynamic wave processes
that take place upon the bursting of the "diaphragm"
in a shock tube are described by means of a time-distance diagram on Fig. 1.1 (see page 16).The scales of
coordinates are there sufficiently "course", so that
the initial processes of shock formation that occur at
the time when the diaphragm is burst are not "visible".
What one can see then is the wave system obtained as

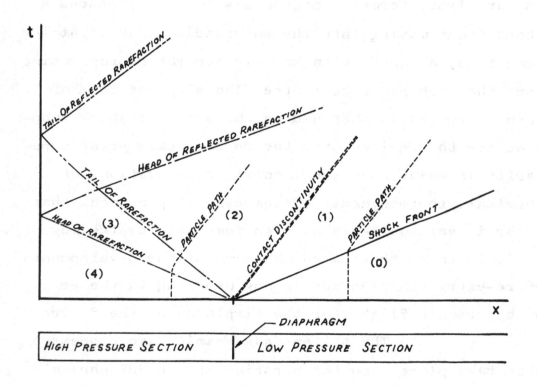

Fig. 1.1. Gasdynamic wave processes in a conventional shock tube.
The indicated regions are: (0) undisturbed low pressure gas;
(1) gas overtaken by the shock front;
(2) gas which passed through the rarefaction wave;
(3) the rarefaction wave;
(4) undisturbed high pressure gas.

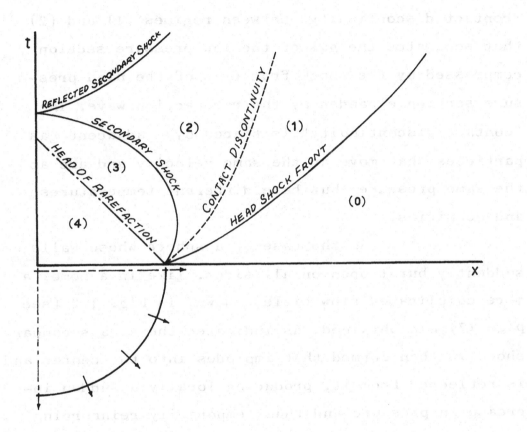

Fig. 1.2. Gasdynamic wave processes in a spherical shock tube.
The indicated regions are: (0) undisturbed atmosphere;
(1) gas overtaken by the head shock front;
(2) gas initially in the bomb that passed through the secondary shock;
(3) expanding gas;
(4) undisturbed gas in the bomb.

the result of the bursting of the wall rather than those associated with this event. Of particular interest in this connection is the formation of a so-called "contact discontinuity" between regimes (1) and (2) that separates the gas of the low pressure section compressed by the shock from that of the high pressure section expanded by the rarefaction wave. The "contact discontinuity" is traced by adjacent gas particles that move at the same velocity and are at the same pressure, but have different temperatures and densities.

In the case of a sphere whose walls suddently burst open on all sides, like in a bomb, a more complicated flow field, shown, in Fig. 1.2 (see page 17), is obtained. As indicated there, a secondary shock is then formed that implodes into the center and is reflected from it, producing locally a sudden increase in pressure and thus temporarily reinforcing the explosion process after its initial fast decay.

As it should be evident by now, however, all the explosion processes described so far are essentially those of decaying blast waves, the energy having been supplied when the source was pressurized while throughout the explosion the total energy contained in all of the gas particles undergoing this

process (that is, the sum of their internal thermal
and kinetic energies) remains invariant. If, however,
the energy for explosion is derived from the breaking
of inter-molecular or inter-atomic bonds, as in the
case of a chemical or nuclear explosion, there is a
significant period of time, albeit very short, when
the explosive process is actually "born," that is,
the initial period when the generation of the non-
steady motion of the medium is intimately related
with the exothermic processes of energy release in
the reacting medium. If, moreover, the wave front of
the explosion propagates through a medium that is
capable of reacting sufficiently fast, it can move at
a constant velocity-a process known as detonation.
Such a wave is then considered to be "self-sustained"
and thus is has also a "life" of its own, in contrast
to the blast waves obtained from high pressure vessels
which, as pointed out obove, are entirely in the
process of "lifeless" decay.

Thus, the only explosions that have
"birth and life" are those produced by chemical or
nuclear reactions. In what follows we shall concen-
trate on just chemical reactions, since only these
systems have been studied so far, while, although by
no means impossible, similar studies with nuclear reac-

tions have not yet been performed.

1. 3. Gasdynamics of Explosions.

The birth and life of explosions have
two outstanding features : the energy is deposited in
the medium at a high specific power level (that is the
amount of energy deposited per unit mass in a unit of
time is high) and the medium where this process takes
place is set thereby in a nonsteady motion. The latter,
in turn, influences the rate at which the energy is
evolved. It is the study of the interrelationship between these two processes that constitute the principal topic of <u>Gasdynamics of Explosions</u>, a new branch
of Gasdynamics that today addresses itself principally to the question : what is exactly the mechanism
of the evolution of the driving power (or propulsive
force) for the generation of explosions (or the growth
of blast waves) that is produced as a result of the
energy released by fast exothermic reactions as those
occurring in explosive gas mixtures. In what follows,
the experimental techniques used in this field of
study and their most prominent results will be described.

The laboratory apparatus consists usual-
ly of a closed vessel where the test gases are initial-

ly maintained at a specified homogeneous state at
rest, and the gasdynamic flow field is generated either by the shuttering of a diaphragm between the
test vessel and a high pressure reservoir to which
it is for this purpose connected (shock tube), or by
igniting the medium when the test gas is an explosive
mixture (detonation tube), or by firing a projectile
through the vessel (ballistic range), or by a combi-
nation of these devices. For the observation of the
flow field and the measurement of pressure pulses,
high frequency response or, what amounts to the same
thing, short rise-time instruments are necessary.

With respect to the time resolution
required to observe the birth and life of explosions,
such instruments became available only quite recently,
thanks primarily to the progress in solid state elec-
tronics and laser technology. The overall duration of
the most significant period of the explosion process
that bears directly upon its generation is, under
normal circumstances, less than a microsecond (10^{-6}
sec) and it takes, therefore, an instrument with a
resolving power in nanoseconds (10^{-9} sec) to observe
its evolution. It is this capability then that has
been provided to science just recently by lasers that
can yield repetitively (i.e., stroboscopically) light

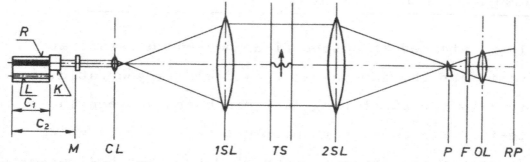

Fig. 1.3. Stroboscopic laser schlieren system.

(a) Optical layout

C_1 light pump cavity, C_2 laser cavity, L light pump, R ruby rod, K Kerr cell for Q-spoiling modulation of laser output, M lase mirror, CL condenser lens, 1 SL front schlieren lens (or mirror), TS test section, 2 SL back schlieren lens (or mirror), P schlieren aperture (quartz prism), F polarizing filter, OL objective lens, RP receptor plane.

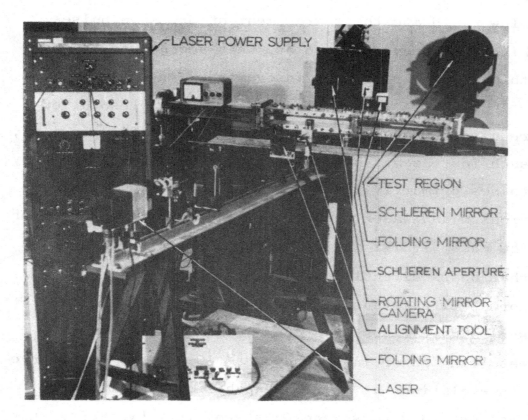

Fig. 1.3. Stroboscopic laser schlieren system.

(b) Experimental set - up.

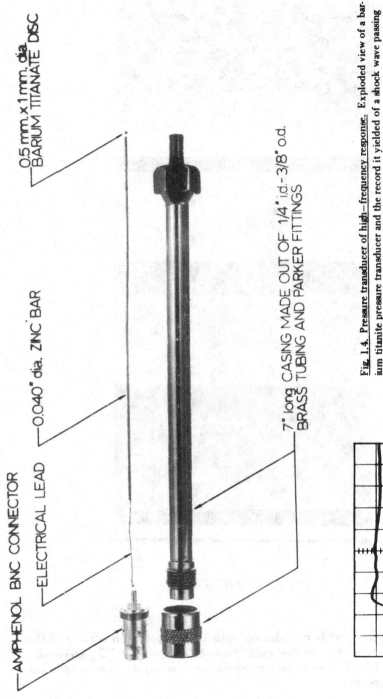

0.5 mm. x 1 mm. dia.
BARIUM TITANATE DISC

0.040" dia. ZINC BAR

AMPHENOL BNC CONNECTOR

ELECTRICAL LEAD

7" long CASING MADE OUT OF 1/4" i.d. - 3/8" o.d.
BRASS TUBING AND PARKER FITTINGS

Fig. 1.4. Pressure transducer of high–frequency response. Exploded view of a barium titanite pressure transducer and the record it yielded of a shock wave passing across its face at a velocity of 530 m/sec. The medium was nitrogen at an initial atmospheric pressure and room temperature, so that the shock Mach number was 1.502 and the recorded pressure step 22.4 psi. The record was obtained by means of a Tektronic Type 533A oscilloscope with the horizontal sweep set at 2 μsec/cm and the vertical deflection at 0.02 volts/cm.

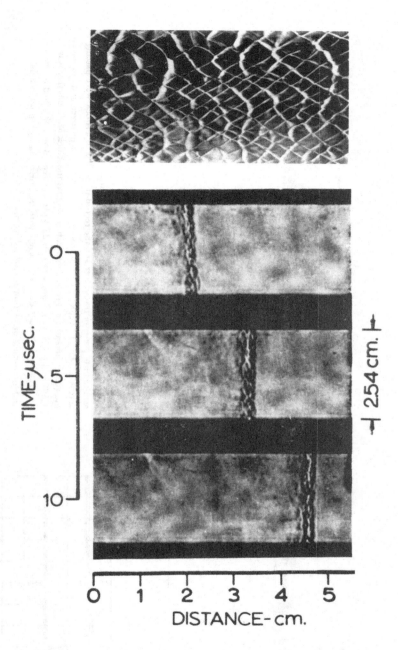

Fig. 1.5.   Stroboscopic laser-schlieren photographs of a detonation wave and the soot imprint recorded on the wall. The medium was a $4H_2 + 3O_2$ mixture contained in a $1'' \times 1\text{-}1/2''$ cross-section tube and maintained initially at a pressure of 87.3 mm Hg and room temperature.

pulses of only several nanoseconds in half width. To
realize the significance of such short time resolution
one should note that, with respect to the number of
"bits of information", the time period of a microsecond has become today equivalent to a quarter of an
hour of experimental time when the frequency of scientific observations was at a level of a kilocycle per
second, that, indeed, has been the case not too long
a time ago. The most direct application of such light
sources is obtained by means of the schlieren system-
an optical technique based on the exploitation of
light refraction. Specifically, as shown in Fig. 1.3
(see page 22), one passes for this purpose through the
test section a parallel beam of light which refracts
at the fronts of appreciable density gradients (in the
direction normal to the optical axis,) that are characteristic of shock and combustion waves. The refracted
light is screened out by an appropriately placed
"schlieren apperture" and the recorded image of the
flow field retains thus traces of these fronts. Since
the cold medium ahead of the wave and the hot medium
behind are devoid of density gradients, they appear on
the record as unobstructed regions. A more detailed
description of such systems as that represented by

Fig. 1.3 can be found in Refs. 1 and 2*). The salient
features of the concomitant instrumentation have been
described most comprehensively by Soloukhin (3) . Of
this, the most important for gasdynamic studies of
explosions is a pressure transducer of high-frequency
response. An example of such an instrument and its
performance is given by Fig. 1.4 (see page 23).

As to the results of such an observa-
tional technique, let us take a look at Fig. 1.5 (see
page 24) representing a sequence of schlieren photographs taken at a frequency of 200 kilocycles per
second (that is, at 5 microsecond intervals) of a
detonation wave-a wave system known for a long time
(actually since 1882) that, according to the classical
concept still prevalent today, is thought of as a
plane shock followed by a combustion wave. The photographs of Fig. 1.5 reveal that nothing of this sort
actually takes place ! At it appears there, the wave
is essentially cellular in structure and its front
consists, in fact, of a system of intersecting cusps,
as it is schematically described by Fig. 1.6 (see page
27).

---

*) Numbers in parentheses denote references listed at the end
of the chapter.

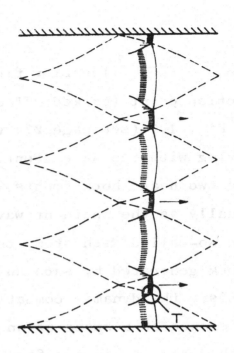

Fig. 1.6. Schematic diagram of a detonation wave.
The solid lines represent curved intersecting shock
fronts. The shaded regimes are the combustion
zones. The broken lines depict the traces of the
wave intersection points. The arrow indicate the
positions where the wave system may receive the
propulsive drive for its sustenance--the source of
its "life".

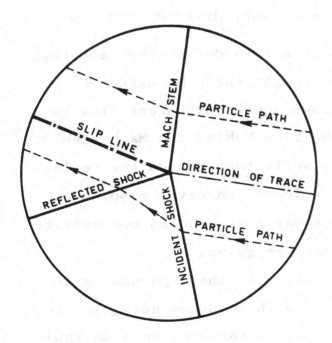

Fig. 1.7. Triple shock intersection.
The flow field is presented from the
point of an observer fixed with respect to the wave front. The direction
of the trace such a triple point intersection would make is parallel to the
particle path incident upon the wave
system as "seen" by this observer.

The flow field existing at each inter-section point (marked "T" on the sketch) is depicted on Fig. 1.7 (see page 27) with respect to an observer moving with the wave front. As evident there, besides the two head shock fronts, the weaker one referred to usually as the incident wave, while the stronger as the so-called Mach stem, there is also a reflected shock generated by such an intersection in order to satisfy the dynamic compatibility requirement for the existence of a state of uniform pressure and particle path direction in the flow regime behind the two fronts. This regime is, moreover, divided into two parts by a slip-line that arises due to the fact that the gas particles were brought there by different routes and have, therefore, quite different flow ve-locity at its sides, the flow behind the Mach stem being locally subsonic, while that behind the reflec-ted shock is supersonic. This then gives rise to a considerable amount of shear that creates concentrat-ed vortex of high temperature gases.

The net effect of these phenomena is quite spectacular : since such a vortex acts, in ef-fect, as a rotating stylus, it can etch on a suitable

material the trace of its path, and the detonation
front is rendered thus the Biblical property of
being able to write on the walls ! In the laboratory
experiment such writing is obtained by having the
wall of the detonation tube covered with a thin
layer of carbon soot. Usually this is done by coating
a thin plastic sheet with a deposit of a smoky flame
of a match or a kerosene lamp, and then gluing this
sheet to the inner surface of the tube. Such a record
of the detonation wave whose schlieren photographs
are presented on Fig. 1.5 is displayed on this
figure above the photographs.

      The soot technique can be combined with
a stroboscopic laser system by having one of the
glass walls of the test section coated directly with
the layer of soot and photographing the back lighted
test section with the parallel beam of the optical
system, but without the schlieren aperture. The
record obtained in this manner consists then of
direct shadow photographs of the wave pattern and of
the traces of the triple points along the wall. This
is demonstrated on Fig. 1.8 (see page 31) where the
superposition of the two types of records of Fig.1.5
is quite evident.

      For a more detailed study the size of

the cells can be increased. This is done by lowering
the power density level of energy deposition- a procedure which can be easily accomplished either by de-
creasing the initial pressure of the explosive gas
mixture or by adding to it some diluent gas, such as
nitrogen or argon. Instead of the numerous multi-cell
structure, one can obtain then a double cell pattern,
as that represented by Fig. 1.9 (see page 32) or, in
the limit, a single mode as shown on Fig. 1.10 (see
page 33). The stationary traces on these records are
the paths traversed by triple points, while those
moving from left to right are the wave fronts. The
photographic records of Figs. 1.9 and 1.10 have indeed caught the wave fronts in the act of "writing on
the walls" and they provide, in effect, the experimental proof of the mechanism of this process, as described here earlier, since the traces appear there
clearly as they are being etched by the sharp intersection points of the shock fronts.

This is not all. The regime of high
temperature that is concentrated by the action of
the vortex and reinforced by collisions between the
triple points is associated also with luminosity
which is considerably higher than that of the rest of
the flow field. One can then record the traces of the

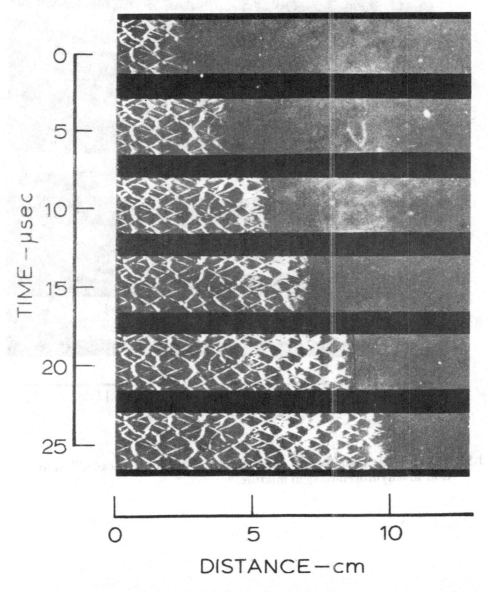

Fig. 1.8. Stroboscopic laser—shadow photographs of detonation in a hydrogen—oxygen mixture.

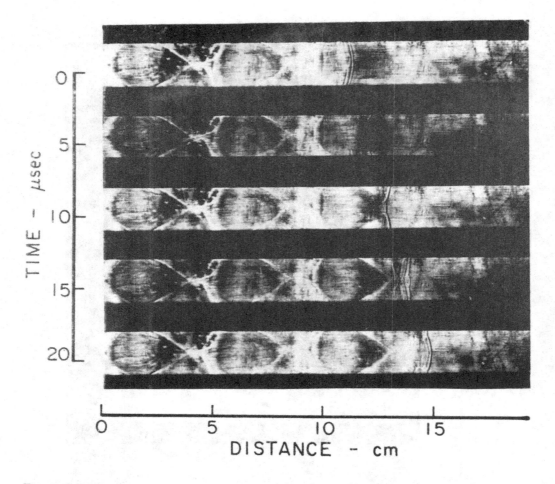

Fig. 1.9. Stroboscopic laser shadow photographs of a "double headed" detonation in a hydrogen-oxygen mixture.

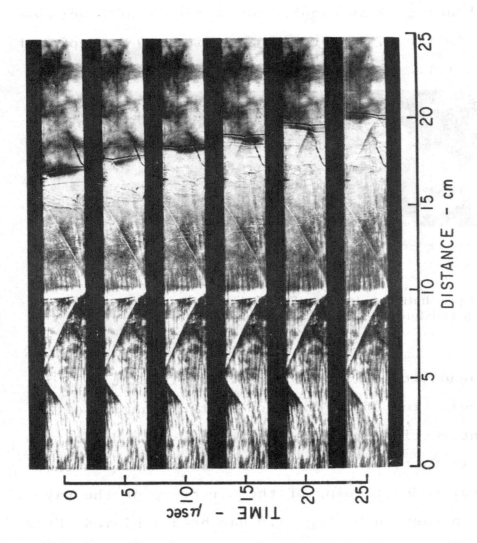

Fig. 1.10 Stoboscopic laser shadow photographs of a "single headed" detonation in a hydrogen-oxygen mixture.

triple points by open shutter photography, similarly
as one would take a snapshot of the paths of automobile headlights at night. Such a record of a detonation, similar in type to that of Fig. 1.9 is depicted
on Fig. 1.11.

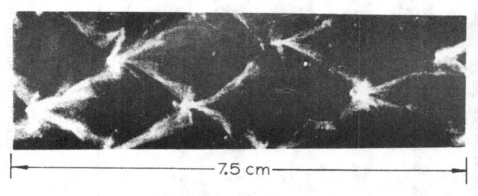

—————————————— 7.5 cm——————————————

Fig.1.11   Open shutter photograph of detonation in an equimolar acetylene-oxygen mixture in a 1/8" x 1" cross- section duct.

        These simple techniques provide valua-
ble diagnostic tolls that are of significant help in
the acquisition of information on the dynamics of
wave interaction processes associated with detonation.
It is, in fact, as a result of these techniques that
the novel understanding of the structure of the wave
front represented by Fig. 1.6 has been achieved. From
the open shutter photograph of Fig. 1.11 it appears
that the intensity of the process-or the power density
of energy release-reaches high peaks at the collision
points, followed by a decay of the wave system between

the subsequent collisions. Thus, as manifested by
the dark patches along the traces, the process becomes actually extinguished before it gets revived
again by the next collision(4) . Hence, as in true
living matter, the wave has not only a cellular structure, but the life span of each cell is comparatively
short, so that all the elements of the system,
throughout its existence, undergo periodic cycles of
decay and restoration, while the whole is in the
state of continuous progress as a self-sustained or
"living" phenomenon.

        Having become acquainted with the cha-
racteristic features of the life of the explosion,
let us now delve into its birth. As it has already
become evident from the foregoing, the necessary
ingredient for this purpose is the deposition of a
certain amount of energy into the explosive gas mixture at a sufficiently high rate. In an essentially
one-dimensional flow field of a slender tube, it is
sufficient for this purpose to ignite the mixture at
one end by a tiny spark. What happens then in a
stoichiometric hydrogen-oxygen mixture, maintained
initially at atmospheric pressure and temperature,
is shown by the stroboscopic sequence of schlieren
photographs in Fig. 1.12.

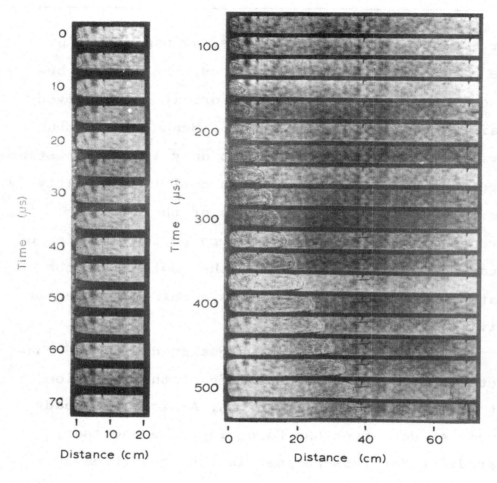

Fig. 1.12(a)   Fig. 1.12(b)

Fig. 1.12 Stoboscopic laser schlieren photographs of flame acceleration in a
          hydrogen-oxygen mixture ignited by a spark neat the closed end of a
          1" x 1 − 1/2 cross-section tube.

(a) Ignition ; blast wave from a 1 millijoule spark by a laminar flame kernel

(b) Generation of a shock (see last frame) by accelerating flame.

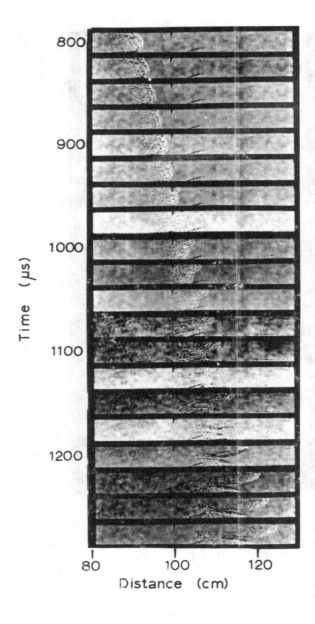

Fig. 1.12 (c) Transition to turbulent flame.

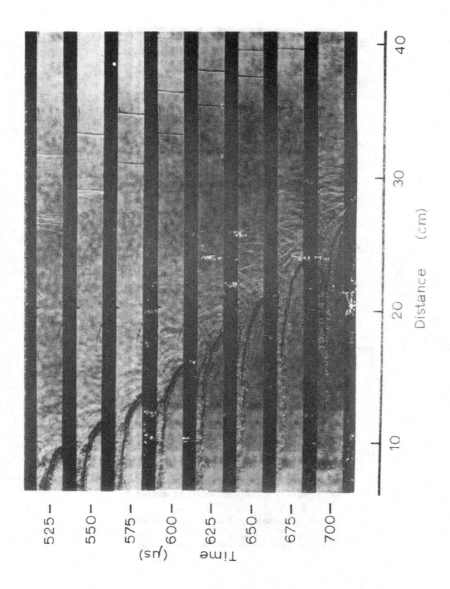

Fig. 1.12 (d) Later stages of the process manisfested by the generation of intense pressure waves immediately ahead of the turbulent flame.

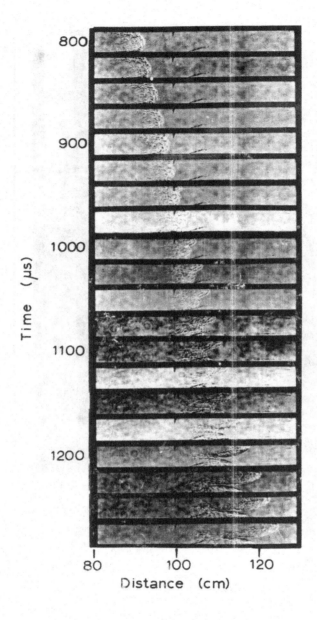

Fig. 1.12 (c) Transition to turbulent flame.

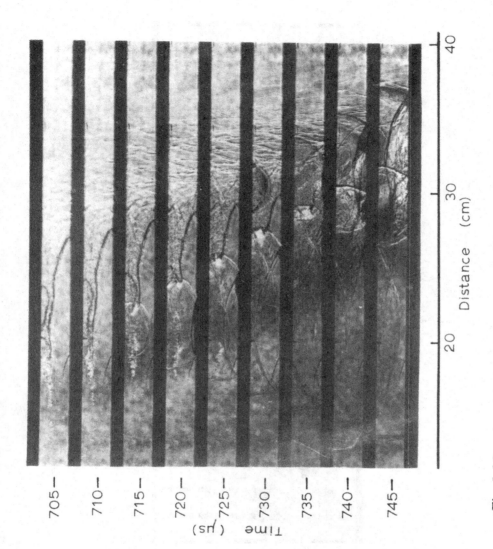

Fig. 1.12 (d) Later stages of the process manisfested by the generation of intense pressure waves immediately ahead of the turbulent flame.

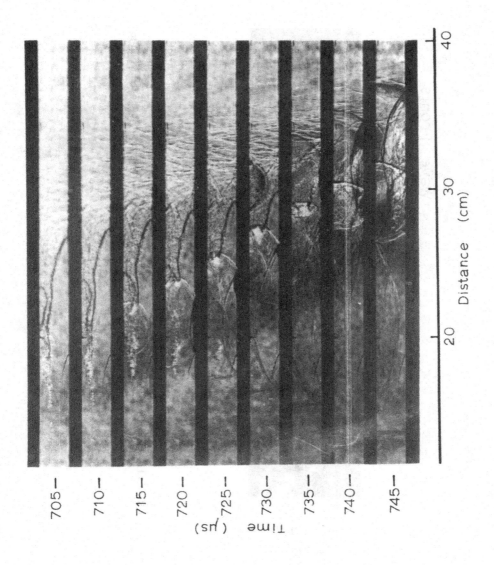

Fig. 1.12 (e) Onset of detonation as a consequence of secondary explosions in the exploding gas.

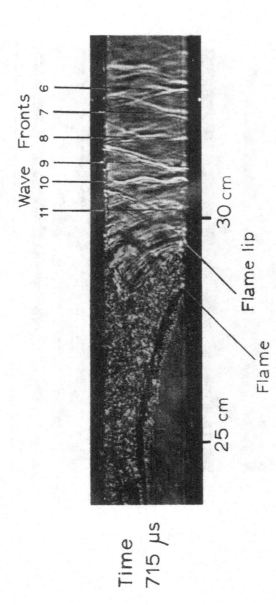

Fig. 1.12 (f) Enlargement of pressure waves ahead of the turbulent flame prior to the transition to detonation.

The flame starts first as a smooth-surfaced spheroid,
Fig. 1.12a. As its front becomes wrinkled, it acceler
ates, generating pressure waves that coalesce into a
shock front as it can be seen at the bottom of Fig.
1.12b. Later, as shown on the subsequent sequence of
photographs, Fig. 1.12c, the flame undergoes a transformation into a turbulent structure, while its surface
acquires a characteristic caved-in, "tulip-shaped"
form. This transformation is again associated with a
more intense generation of pressure waves. These waves
merge forming shock fronts, as shown on the first frame
on Fig. 1.12d representing a much later stage of the
phenomenon. The process becomes thereby greatly reinforced, yielding a train of intense pressure waves that
emanate from the-by ten-highly turbulent flame, as it
appears in the last frames of this figure. This, in
turn, leads to an explosion in the exploding medium as
seen on Fig. 1.12e, depicting the events following directly those recorded by Fig. 1.12d, and representing,
in effect, the last stages in the birth of a detonation
wave. Actually, there have been two such "explosions
in the explosion" recorded on this figure, one first
visible at 715μsec on the time scale and the location
corresponding to about 20 cm on the distance scale, the
second appearing first at 730μsec and at approximately

30 cm. In this particular case both the secondary explosions produced detonation fronts which are seen
propagating to the right at the lower frames of Fig.
1.12e. The detonation due to the first explosion will
have to decay later, of course, since the other will
consume all the unburned mixture, leaving just essentially inert products behind. The details of the pressure field generated by the turbulent flame are shown
by Fig. 1.12f, and the time-space diagram of the whole
process, based on all the records of Fig. 1.12, is presented on Fig. 1.13.

In an unconfined space, the initiation
of the detonation seems to be quite a sensitive function of the amount of ignition energy. This has been
demonstrated recently by an interesting set of records
obtained by J.H. Lee and his associates at McGill University in Montreal, using a laser pulse for ignition
to initiate the process practically at a point (5).
In the sequence of schlieren photographs, Fig. 1.14a,
(see page 45) this energy was evidently insufficient
to promote detonation since the combustion front is
lagging behind the shock, and such a wave system is
therefore in the state of decay,that is,the reaction
is decoupled from the shock front and, from then on,

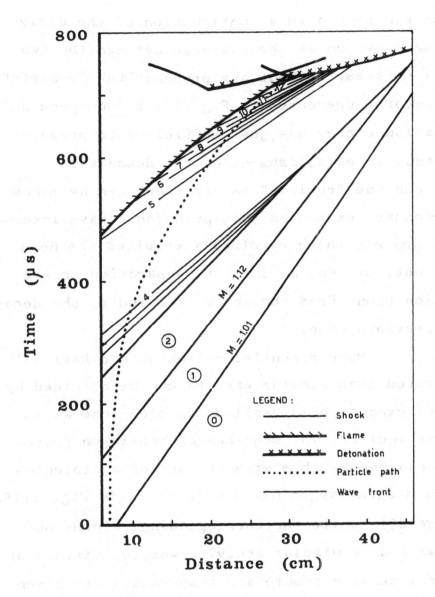

Fig.1.13 Time-distance diagram of wave processes associated with flame acceleration in an explosive gas. Wave front traces were deduced from Fig. 1.12, the numbers corresponding to the fronts indicated on Fig. 1.12f. The dotted line denotes the particle path that leads to the center of the "explosion in the explosion" that triggered the transition to detonation.

all that can happen is a continuation of the decay
of the wave system as the distance between the two
fronts increases, without any prospect for "rebirth".
In the records presented in Fig. 1.14b (see page 46)
the ignition energy was just sufficient to promote
an eventual re-establishment of the detonation
process and the "rebirth" is attained then by means
of a secondary explosion that produced a wave intersection process which eventually engulfed the head
shock front, preventing thus the separation of the
combustion front from the shock that led to the decay
in the previous case.

　　　More recently, Soloukhin and Ragland[6]
demonstrated that similar effects can be obtained by
using the exhaust of a small detonation tube as an
explosive igniter. The sequence of schlieren photographs obtained by them with the use of a stoichiometric hydrogen-oxygen mixture is shown in Fig. 1.15a
(see page 49), while the corresponding records obtained with an equimolar acetylene-oxygen mixture at
the same initial pressure and temperature are given
by Fig. 1.15b (see page 50). The process recorded on
Fig. 1.15a corresponds evidently to that of Fig.1.14a
while the records of Fig. 1.15b are in effect similar to those of Fig. 1.14b.

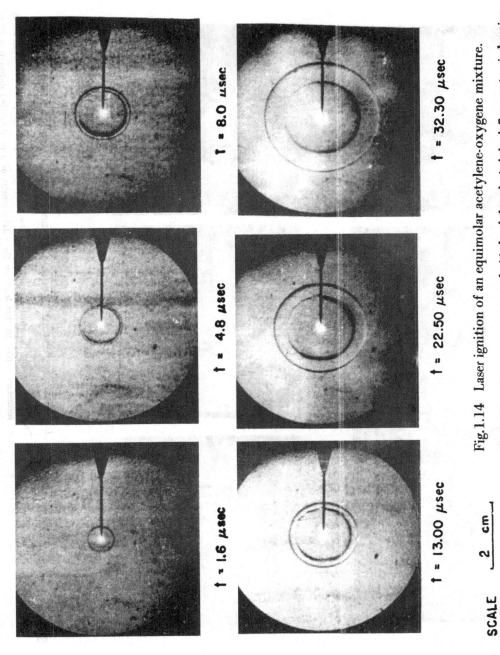

t = 1.6 μsec   t = 4.8 μsec   t = 8.0 μsec

t = 13.00 μsec   t = 22.50 μsec   t = 32.30 μsec

SCALE   2 cm

Fig.1.14   Laser ignition of an equimolar acetylene-oxygene mixture.

(a)   The ignition energy was near subcritical and the spherical shock-flame system is decaying.

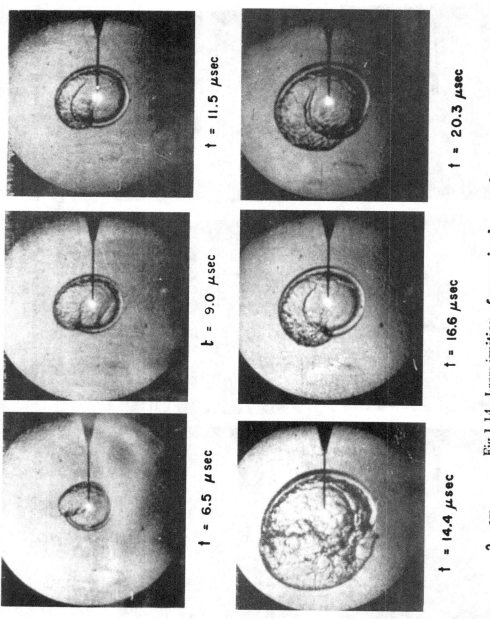

t = 6.5 μsec   t = 9.0 μsec   t = 11.5 μsec

t = 14.4 μsec   t = 16.6 μsec   t = 20.3 μsec

SCALE   2 cm

Fig.1.14   Laser ignition of an equimolar acetylene-oxygene mixture.

(b)   The ignition energy was just overcritical and the shock-flame system is prevented from decay by secondary explosion that leads to the establishment of a multiheaded detonation front.

To sum up, the most significant discovery made recently concerning the mechanism of explosions driven by chemical reactions (that is detonation waves) is the realization that the propagation of the wave is essentially due to processes occuring at a distinct number of reaction centers that are extremely small in size (e.g., less than 1 mm for a hydrogen-oxygen system initially at normal pressure and temperature) and where the driving power is being actually evolved in an extremely short period of time (less than 1 microsecond). Except for these centers, the rest of the wave system is in a process of decay. The evolution of the detonation wave is essentially due to the action of the same kind of explosion centers.

1. 4. Technological Significance.

Perhaps the most direct demonstration of the utilitarian nature of the explosion process is represented by the evolution of propulsion systems. In principle the purpose of such systems is the conversion of potential energy into kinetic. The input may be provided in a variety of forms, depending on how the potential energy is stored (e.g., thermal, chemical, electrical, nuclear or gravitational) ; the

output, however, is, in kind, always the same : the
kinetic energy of the translational motion of the
vehicle. In the case of jet propulsion systems this
train of throught can be simplified by restricting
one's attention just to the motion of the working
substance itself, yielding the exhaust jet. Since
the potential energy has been also stored in this
substance, the tracing of the processes of energy
conversion that occur in a given system is then quite
straightforward.

The evolution of propulsion is characterized by the drive to pack more and more power into
the working substance, without bringing about a
drastic change in the overall size of the system,
or, in other words, by the increase in the power
density of energy deposition. This is illustrated by
Fig. 1.16 (see page 52) which displays the spectrum
of power density deposited in the working substances
of various prime-movers.

Thus in a steam plant the power density
of energy deposition into water is of an order of
1 kilowatt per liter ; in a cylinder of an internal
combustion engine this can reach a level of 100 kilo-
watts per liter, while in a turbojet combustor it can
be as high as 1 magawatt per liter. In a high output

μsec

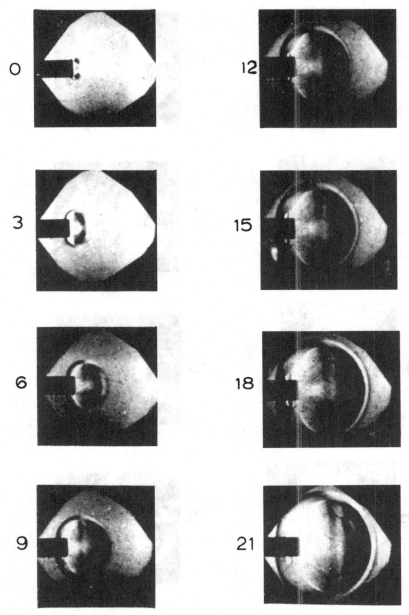

Fig.1.15 Ignition of an explosive gas mixture by a detonation wave. The small detonation tube used as igniter is 10 mm i.d. and 13 mm o.d. ; time interval between frames : 3 usecs. (a) Subcritical ignition obtained in the case of an equimolar hydrogen-oxygen mixture at an initial pressure of 0.07 atm.

μsec

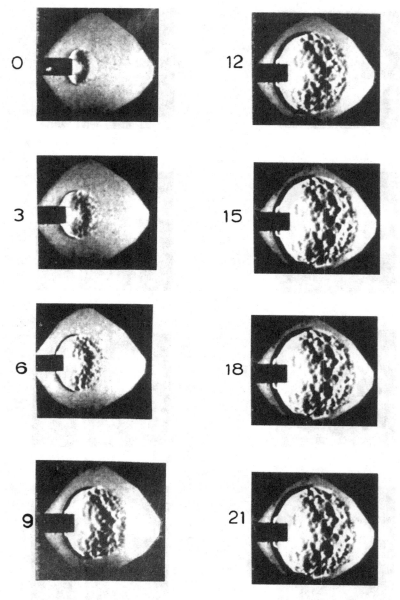

Fig. 1.15 Ignition of an explosive gas mixture by a detonation wave. The small detonation tube used as igniter is 10 mm i.d. and 13 mm o.d. ; time interval between frames : 3 μsecs. (b) Supercritical ignition obtained in the case of an equimolar acetylene-oxygen mixture at the same initial pressure as (a).

rocket thrust chamber one achieves in this respect
tens of megawatts per liter, the hydrogen-oxygen system representing today the most advanced state of art
with a power density level approaching 100 megawatts
per liter. Figure 1.16 demonstrates also the power
densities realizable in gaseous explosions indicating
their leading position with respect to future developments of prime-movers for propulsion.

Another way to describe this trend is
by considering just the performance range of rocket
propulsion systems. The usual form of displaying this
is by means of a diagram, or map, in the plane of
thrust to mass ratio, or vehicle acceleration, used
as the ordinate, and the so-called specific impulse,
or jet velocity,forming the abscissa. Such a map is
shown in Fig. 1.17 (see page 52), where, it should
be noted,the scale of the abscissae extends all the
way up to the velocity of light.

The product of the two coordinates in
this diagram is proportional to the specific power
output of the propulsion   system,

(1.1) $$P_s = \frac{VF}{2M} = \frac{1}{2} I_s a$$

where V represents the jet velocity, F the thrust,
and M the vehicle mass. As demonstrated by the

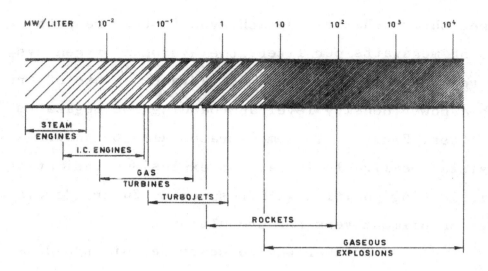

Fig.1.16 Power density spectrum of energy deposition in combustion chambers
of prime movers.

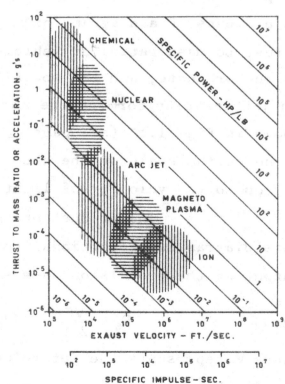

Fig. 1.17. Map of the performance
range of rocket propulsion systems
on the plane of vehicle acceleration
versus specific impulse.

second expression, this can be also expressed in
terms of the product of the so-called specific impulse, I_s , and the vehicle acceleration , a . As a
consequence of the logarithmic scales in Fig. 1.17
the hyperbolae of constant P_s are represented by
straight lines inclined at an angle of 45° to the
axes of the coordinates. It is immediately apparent
from this diagram that attainable performance of all
the systems, that represent there in fact the technological standards of today, is limited by the power
density at a level of about 100 hp/lb_m

      The performance of thrust chambers
alone can be also described by a similar map, as
demonstrated by Fig. 1.18 (see page 54). Here, in
contrast to the previous case, the scale of ordinates
refers to the ratio of the force to the mass of the
working substance, rather than of the thrust to the
mass of the whole vehicle as before, while the scale
of abscissae remains the same. The diagram of Fig.
1.18 represents, in effect, all the mechanical parameters associated with the acceleration of the working substance to the exhaust jet velocity. Hence, in
addition to the specific power lines, it embodies
also lines of constant "effective time"

(1.2)
$$T = \frac{V M}{F} = \frac{V^2}{2 P_s}$$

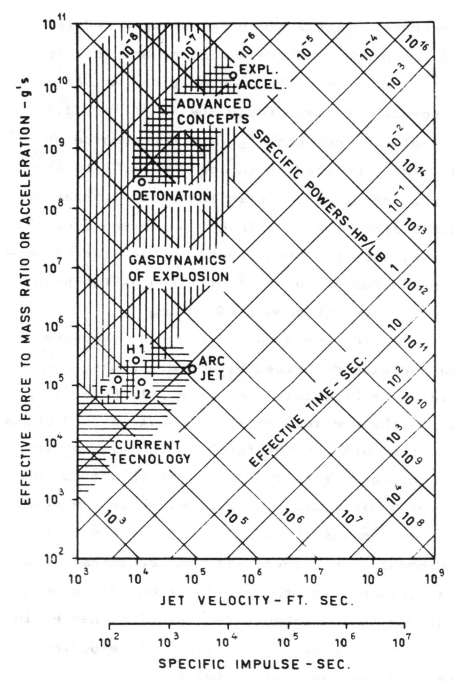

Fig.1.18 Map of the performance range of thrust chambers on the plane of effective force to mass ratio versus jet velocity.

that is, the time required for the acceleration of
the working substance from rest to the final jet
velocity under the action of a constant force.

The range of operating conditions,
representing the standards of CURRENT TECHNOLOGY,
has been delineated on the basis of the performance
parameters of the F1, H1 and J2 engines, and the es-
timated capability of an advanced arc-jet thrustor.
The domain of ADVANCED CONCEPTS is based on the parameters corresponding to the so-called Orian propulsion system based on the use of atom bombs exploded
externally behind the space vehicle, an explosive
accelerator which, by multi-staging, offers the possibility of attaining kinetic energies equivalent to
fusion temperature [7,8] and the detonation wave in a
hydrogen-oxygen mixture whose photographic records
have been shown here in Fig.4. The data on which the
point representing detonation in Fig. 1.18 is based,
are : the energy release of 3 kcal/gm in a wave
3.5 mm thick moving at a velocity of 2.5 km/sec into
a medium at a density of 0.1 mgm/cm^3.

The specific power of energy deposi-
tion, according to current technological standards,
is of the order of million (10^6) hp/lb, while the
effective time is about a millisecond (10^{-3} sec).

The advanced concepts are at a level of trillion
(10^{12}) hp/lb in specific power and a microsecond
(10^{-6} sec) in effective time. The regime labeled
GASDYNAMICS OF EXPLOSIONS indicates the performance
of propulsion systems which can be attained by the
exploitation of this subject of study. It covers the
range of operating conditions, overlapping both the
CURRENT and ADVANCED technologies, and it should be,
therefore, instrumental in bridging the gap between
the two.

1. 5. Future Prospects.

In retrospect, we became acquainted
with the physical nature of the life and birth of
explosion processes, learned something about the manner in which they are studied, and had a glimpse at
some of their fruitful yields, notably in propulsion
devices, notwithstanding our realization of their so
well known destructive power. With this background
it is therefore quite proper to ask what specifically
can be gained from such knowledge in the future.

As it has been already pointed out, ba-
sically we will gain a better understanding of the
dynamics of the processes where energy is deposited
in a medium at a high specific power level. But what

does this mean in terms of specific technological and scientific advances ?

As this time the reader may have formed visions of propulsion systems based on the use of detonative high explosive media and having a physical appearance completely different from the conventional devices used today for this purpose. Although some work on the development of such systems has, indeed, been done,as exemplified by the concept of the Orian project mentioned here earlier in connection with Fig. 1.18,and a variety of proposals have been advanced for the design of rocket thrust chambers based on the use of detonation waves, this is certainly not the point here.

As implied by the diagram of Fig.1.18; it is not so much the actual attainment of the ADVANCED CONCEPTS, as the bridging of the gap between them and the CURRENT TECHNOLOGY that is the proper province of GASDYNAMICS OF EXPLOSIONS. The main point advocated here is that, by understanding better the gasdynamic processes associated with energy release due to fast chemical reactions, we should be able to improve the performance of our engines in one significant way : increase the power density level at which the energy release takes place.

What would then be the specific gains?
Let us examine this question first with reference to
the conventional propulsion systems that we know today. As a preamble. one should realize that incipient
explosions are *ae facto* present in practically any
modern combustion system used for propulsion. from an
automobile engine to the thrust chamber of the Saturn
rocket booster where the upper stages use actually
the same hydrogen-oxygen system that provided us here
with most of the examples of explosion phenomena.
Since the dynamic effects of explosive combustion are
destructive. the process of energy conversion is con-
trolled there in such a way that any tendency for explosion is severely restricted. As the detailed nature
of the explosion process is not known, however, not
only is the limit to which this tendency should be
contained unclearly defined, but even the sense of
direction for future evolution in this respect is
just vaguely comprehended.

To be more specific, for the internal
combustion engines of an automobile, for instance, the
importance of high compression ratio is well established. The phenomenon that imposes a limit on the
compression ratio is referred to as "knock". The abil-
ity of a given fuel to operate at high compression

ratios without "knocking" is measured in terms of a
vague quantity known as the octane number but-and
this is significant-the gasdynamic mechanism of
knock is _not known_. Is it then not reasonable to expect that if its details were known, the composition
of fuels could be altered to permit even higher compression ratio than those used today ?

Should we succeed in this respect,
then as a concomitant advantage, one should be able
to accomplish the combustion process in a smaller
volume and shorter time than it is done today, use
leaner mixtures leading to the production of more
inert exhaust gases and eliminating thus most of the
sources of air pollution due to car exhaust.

As far as rocket engines are concerned,
although most of the dynamic processes of energy conversion in a compressible medium that have been described here do indeed occur in their thrust chambers,
normally they are neglected because of the relatively
short period of time and small size of the explosion
centers. Thus, for obvious reasons, a much larger importance is attached to the parameters describing the
overall behavior of the system, rather than to the
purely local effects of tiny, short-lived "hot spots".
However, whereas it is indeed true that one's first

goal should be the understanding of the overall behavior, one should not overlook the importance of know-
ing the details of the processes occuring at the heart
of the system, even if they are so short in existence
and occupy such small volumes in space, especially in
view of the fact that these processes are occuring at
such high frequencies. After all, it is only by this
knowledge that one can expect to solve, once and for
all, most of the problems associated with combustion
instability that have been plaguing the design of
large rockets right from the beginning of their development and are still imposing today a definite limit
on their performance. Moreover, it is quite reasonable to expect that the attainment of higher power
densities in thrust chambers would permit their oper-
ation at significantly higher pressures and at the
same time with much higher velocities of flow and,
consequently, smaller nozzle throat contraction ratios, thus contributing towards the improvement in
their performance.

        As to the scientific advances, one
should realize that the dynamic properties of explo-
sions-the principal subject of Gasdynamics of Explo-
sions-are governed, on one hand, by the kinetic processes of fast chemical reactions and, on the other,

by the gasdynamic effects of wave phenomena. The
chemical events of interest here are, as a rule,
associated with chain reactions which proceed essentially in three stages : at first they start with an
induction period during which molecules of reactants
are broken up, building up a supply of active species
(chain carriers). This takes a relatively long time,
typically several microseconds in duration, since
at the beginning there is practically zero concentra-
tion of such species. Then, there is a relatively
short period-one microsecond or less-of the actual
heat release, the most active life of the phenomenon
when most of the energy is deposited in the medium
producing more chemical bonds to form new molecules
and yielding higher kinetic energy of translational
motion that is manifested in the form of pressure
waves. This is normally followed by a very long
period-extending to some milleseconds in duration-of
approach to equilibrium during which all the radicals
acting as chain carriers are gradually deactivated
until only stable or "frozen" species remain in the
system.

Currently it is believed that in the
simplest cases, such as the hydrogen-oxygen system,
the induction process is reaching a satisfactory

level of understanding,but the recombination processes
are still vastly unexplored. Moreover, their systematic
study-although of great importance-will not be, by itself, sufficient. This is due to the fact that there
is such a variety of elementary reconstitution steps
involved, most associated with principally three body
collisions, that even if all of these were known, it
will not be at all simple, if not quite impossible, to
deduce from this some meaningful information on the
overall effects. In this respect then the study of the
"births and lifes of explosions"-where only these overall effects are manifested-should be of particular
value. Moreover, it is only the explosion process that
provides not only a rapid heating to stimulate the
reaction, but also an almost equally rapid cooling
associated with rarefaction-the basic ingredient of a
blast wave-that may quench the products of the highest
chemical activity and prevent the decay that would
have taken place if they were allowed to approach
slowly the final equilibrium state. This property of
the explosion process could therefore be exploited
for the production of new chemical species.

        Now, from the gasdynamic point of view,
the wave phenomena are characterized by the action of
shock fronts. These are associated with a significant

loss in available energy that in propulsion devices
is expressed in terms of the loss in the stagnation
pressure that leads to a decrease in the kinetic
energy of the jet-the useful product of the system.
For a given maximum pressure peak of the wave, the
worst in this respect is a single shock front. If
there is more than one shock, the losses in available
energy are diminished, so that in the limit, with an
infinite number of shocks, they become negligible,
and the process acquires the character of a thermody-
namically optimal, i.e., reversible, change of state.
The study of explosion processes reveals that, indeed,
they are associated not with one but with a multitude
of shocks. This suggests then the possibility of gasdynamic means for the improvement of the performance
of propulsion systems. By exploiting the knowledge
of explosion processes one should be able, in principle, to affect the flow field in such a manner that
it would contain a maximum number of shock fronts,
rather than letting them coalesce into a single powerful shock wave. The control, performed in this case
by proper geometrical configuration and timing, the
latter attained by some suitable chemical ingredients
combined with appropriate pressure and temperature
conditions, would permit the exothermic reactions to

occur only at some distinct "spots", causing the
break-up of shock fronts into smaller elements, so
that, eventually, the flow field would become essentially continuous and devoid of any irreversible
effects due to gasdynamic discontinuities.

Thus one is led to conclude that, as
far as the future of energy conversion systems is
concerned, especially those used for propulsion, we
can be either satisfied with the present status of
technology, and expect progress only by the refine-
ment of existing devices, without altering drastical-
ly their power characteristics, or we may strive for
future progress. Should we favor the latter, it seems,
indeed, that the only way we can enhance future developments is by advancing our knowledge of the Gasdynamics of Explosions, that is, by acquiring a better
understanding of the "birth and life of the explosion
process", and thus lay down the foundations for the
development of systems where energy conversion is accomplished at a much higher power density level than
that possible according to current standards.

References.

1. Oppenheim, A. K., Urtiew, P. A., and Weinberg, F.J., "On the Use of Laser Light Sources in Schlieren-Interferometer Systems," Proc. Roy. Soc., A291, 279-290, 1966.

2. Hecht, G. J., Stell, G. B., and Oppenheim, A. K.; "High-Speed Stroboscopic Photography Using a Kerr Cell Modulated Laser Light Source," ISA Transactions, 5, 2, 133-138, April 1966.

3. Soloukhin, R. E., Udarnyi Volny i Detonastsii v Gazakh (Shock Waves and Detonation in Gases), Gos. Izd. Fiz. Mat. Literatury, 175 pp. Moscow, 1963 ; (Transl. By B. W. Kuvshinoff, Mono Book Corp., Baltimore, 1966).

4. Lundstrom, E. A. and Oppenheim, A. K., "On the Influence of Non-Steadiness on the Thickness of the Detonation Wave," Proc. Roy. Soc., A310, 463-478, 1969.

5. Bach, G. G., Knystautas, R., and Lee, J.H., "Direct Initiation of Spherical Detonations in Gaseous Explosives," Twelfth Symposium

(International) on Combustion, The Combus
tion Institute, pp. 853-864, Pittsburgh,
Pennsylvania, 1969.

6. Soloukhin, R.I. and Ragland, K.W.,"Ignition Processes in Expanding Detonations," Combustion and Flame,13,3, 295-302, June 1969.

7. Lunc. M., Nowak, H., and Smolenski, D., "Accelera-
   tors for Jets Formed by Shaped Charges,"
   Bulletin de l'Académie Polonaise des Sciences, Série des Sciences, Techniques, 12,
   5, 295-297, 1964.

8. Voitenko, A.L., "Producing High-Velocity Gas Jets,
   "AN SSSR Doklady 158G, pp. 1278-1280,1964.

Chapter 2.
Evolution of Gasdynamics of Explosions.

The Dynamics of Exothermic Processes.

2.1. Introduction.

Recent advances in the knowledge of blast wave phenomena, stemming from the pioneering work of Taylor (1)[*], von Neumann (2) and Sedov (3), and exposed in a number of monographs (3-7), provided us with a new facility for the study of the mutual effects between shock waves and exothermic processes, such as those occurring in explosions or in supersonic flow fields supported by chemical or nuclear reactions. Classically, this branch of science was associated with the studies of detonation waves that, significantly enough, have been initiated by the originator of the shock tube technique, Paul Vielle who, together with Berthelot (8), was credited by Mallard and Le Chatelier (9) with their "discovery".

Throughout the ensuing eighty years

---

[*])Numbers in parentheses denote references listed at the end of the chapter.

of detonation research, however, the gasdynamic investigations of these phenomena were handicapped by the
belief that they can be treated as solely steady flow
processes. This led to such well known concepts as
the von Neumann-Döring Zel'dovich model, according to
which a detonation wave is essentially a shock front
followed by a steady deflagration wave whose effects
can be expressed in terms of a family of Hugoniot
curves with the conversion degree as a parameter, and
it incited, among others, such famous controversies
as that between Kistiakowsky and Hirschfelder (10) concerning the relative influence of transport processes
in such a model. The matter has been thoroughly studied since then by Wood (11), Erpenbeck (12), and many
others (13). In all this work, one must reiterate however, the wave system was considered as one involving
just steady flow conditions.

It is then the realization, suggested
by White (14) and followed by many contemporary workers
in this field (15-18), that the coupling between the
shock wave and the exothermic reaction must, as a
rule, involve non-steady gasdynamic processes, which
ushered a new era in this branch of science. Thus, a
concept has been conceived that, in contrast to the
familiar steady flow situation that exists in most

shock tube experiments where the reaction proceeds
behind the shock as a result of the steady high-temperature regime it provides without affecting its
strength, the feedback effect of the exothermic process must be manifested by the establishment of a
non-steady flow field between the shock and the reaction zone. In principle, this is due to the basical-
ly different mechanisms governing the collision and
relaxation processes in the shock wave, on one side,
and the kinetic processes of the reaction, on the
other, preventing the attainment of steady equilibrium between the two, as it has been for such an amaz-
ingly long time considered plausible.

      Unlike, then, the typical shock tube
experiments, the acquisition of data obtained under
non-steady flow conditions becomes necessary. Since
especially pertinent in this respect is the establishment of non-steady flow fields bounded by shock fronts
that is, by defintion, blast waves, their significance
for such purpose is self-evident. Problems in this
area can be classified into two groups, depending on
whether the shock front is associated with a rarefaction or a compression wave. The first corresponds, in
principle, to the case of a decaying blast wave, and
the second to that of a "piston driven" or amplifying

blast wave. These two categories are reviewed here
in turn, and this is then followed by a resumé of the
cases known so far of tractable experiments where
wave sustenance by reaction has been observed. The
latter, it is of interest to note, were all carried
out, in contrast to the remarks made above and in a
paradoxical contradiction to the great amount of data
available on detonation waves, under essentially
steady flow conditions.

2. 2. Decaying Wave Systems.

The earliest experimental observations
of reaction zones imbedded in decaying blast waves
have probably been made by White (19)who,in an effort
to obtain a "laminar" detonation, provided his square
cross-section shock tube with a wedge, so that the
multi-structured wave had to pass through a rapid
convergence in the cross-section area from which it
emerged on the divergent side in its most elementary
form, namely, as a single triple-wave intersection.
The flow field behind the Mach stem was then, in effect, identical to that of a decaying blast wave. At
the same time, in contrast to the curved Mach stem,
the incident shock was straight, inducing a uniform,
essentially constant state and locally steady flow

field which was used by White for the measurement of
induction times.

A similar situation occurs close to
the limits of detonability, when the detonation wave
is referred to as "marginal". In this case, it has
been demonstrated recently how the concept of a
decaying blast wave can be exploited for the analysis
of the effect of non-steadiness on the structure of
the wave system (20).

The most straightforward technique to
generate decaying blast wave is, of course, by means
of a source which, following the theoretical approach,
deposits a finite amount of energy in a negligible
amount of time at a concentrated spot in the medium.
Meeting such requirements experimentally, however,
poses great difficulties for, besides the undesirable
effects of debris due to the explosive material, the
existence of an imploding shock, which arises if the
source is of finite size in comparison to the total
extent of the blast wave, cannot be neglected and
the flow field cannot be, therefore, regarded as continuous. Treating, then, the interaction effects
between the shock front bounding the blast wave and
the reaction that occurs within it, presents, under
such circumstances, practically insurmountable prob-

lems.

Nonetheless, a notable contribution to the experimental aspects in this field of study has been provided by Glass (21), and Boyer (22), who developed a spherical shock tube for the study of blast wave phenomena in the vicinity of finite size explosion sources, offering at the same time a facility for the investigation of the implosion process, while, with respect to analytical aspects, some significant advances have been made, especially, in connection with numerous investigations concerned with electric sparks and exploding wires, as exemplified by the works of Oshima (23).

Recently, means to produce what can be virtually considered as point explosions under laboratory conditions have been attained by the use of pulsed lasers which are capable of depositing enough energy by radiation-induced breakdown, within a sufficiently small volume of the test gas, so that the resulting blast wave, just a few centimeters from the source, can be already considered as one due to a point explosion that has been initiated in an extremely short time (24). Under such conditions, then, the non-steady flow phenomena are tractable and an experimental apparatus of this kind should be, indeed, regarded as one providing the most promising means for the study of the gasdynamic effects of exothermic processes.

The pioneering work in this field has been done by Lee(25,26) who demonstrated both the experimental and theoretical advantages of this technique in the study of explosively reacting detonative gas mixtures. More recently, Soloukhin and Ragland[27] showed that similar flow fields can be obtained also by the use of a small detonation tube as the explosion source. The open end of the tube is placed for this purpose in the middle of a vessel having a considerably larger cross-section area. Since such a source uses the test gas that generates exactly the same products as the reaction under study, this technique has the advantage of initiating the blast wave without inflicting the flow field with any debris that one has to contend with when the blast wave is generated by means of high explosives, exploding wires or electric sparks.

2. 3. Amplifying Wave Systems.

In spite of the classical theoretical background on piston-generated pressure waves that dates back to the work of Riemann, (28), with such notable contributions as those of Friedrichs (29) on the generation of shocks in simple waves, of Boa Teh Chu (30) on the generation of compression waves by heat source, and of Jones (31) on a similar effect due to accelerating flames, to mention just a few, comparatively little experimental work has been, so far, carried out in this field.

The most prominent in this respect are undoubtedly the various studies concerned with the development of detonation, both following a mild ignition in an explosive gas mixture (32,33) and behind the reflected shock (34-37).

A simple and powerful means for wave amplification is obtained by area contraction, as it occurs in shock convergences or implosions-a subject that is indeed associated with ample theoretical knowledge (see e.g.,Refs. 4 and 7). In this respect, the pioneering experimental work has been done by Perry and Kantrowitz (38), who obtained shock convergence in an essentially cylindrical geometry by the use of a conical field divider placed in the vicinity of the closed end of the shock tube. A more powerful idea for generating implosions by irradiation has been suggested by Daiber, et al (39), who recommended the use of lasers in a two-pulse irradiation experiment, first to generate a blast wave and then, exploiting the fact that most of the mass of the medium embodied by the wave is then concentrated in a shell near its periphery, the second to initiate the implosion.

Of particular significance to the study of amplifying waves are the various investigations involving detonative gas mixtures. In this respect, especially prominent is the work of Lee and his associates (40-44), who developed a good deal of experimental as well as analytical techniques for the study of imploding detonation waves, in both cylindrical and spherical geometry. Associated with just the induction processes in non-steady compression waves is the interesting investigation of Strehlow (45) who used a simple area convergence in a shock tube to attain the non-uniform field. On the other side of the spectrum, concerned with the generation of extremely powerful means, is the unique method, based on the use of a solid high-explosive initiator and a

high pressure detonative gas mixture, that has been
developed by Glass(46,47) to obtain a stable spherical implosion wave for use primarily as a driver for
a hypervelocity launcher. The system provides,however,
also a convenient, reusable facility for conducting
physical experiments at extremely high pressures and
temperatures.

In the meantime, more groundwork for
experimental studies was provided by various analytical investigations. These include the generation of
pressure waves by distributed combustion zones, as
they occur in particle-fueled combustion systems(48,49)
as well as in a neutron irradiated fissile gas (50,51)
and by a fission reaction initiated behind a reflected shock (52). More recently, an analysis has
been completed on the generation of pressure waves
by an explosive reaction center, leading to the
establishment of a direct interrelationship between
the kinetic processes of chemical reaction and the
non-steady gasdynamic effects of the blast wave (53).

2. 4. Sustained Wave Systems.

Although, as pointed out in the introduction, the primary incentive for the study of exothermic reactions by the use of blast waves was

derived from the interest in detonation phenomena
which provide, in fact, classical examples of self-
sustained waves, so far, paradoxically enough, all
attempts to study by their means the mechanism of
self-sustenance have failed. Thus, the exploitation
of the detonation tube per se for the investigation
of the interaction between the gasdynamic and chemo-
kinetic processes has been, up to now, restricted to
the case of marginal detonations that have been al-
luded to here earlier in connection with the decaying
wave systems.

          More successful in this respect was
the exploitation of steady flow fields attained
either in supersonic jets or in ballistic ranges. The
former, started by Gross (54) and Nicholls (55) and
their associates in an attempt to ottain a standing
detonation wave, yielded eventually a facility for
the establishment of a reaction zone in a rarefaction
field behind the Mach disc in a supersonic jet, which
turned out to be especially convenient for the investigation of the induction process (55,56). The latter
is associated with reactions that occur behind a bow
wave in front of a projectile passing through an explosive gas mixture. After the initial investigations
of Ruegg and Dorsey (57) at the National Bureau of

Standards, a number of interesting results have been
obtained by Behrens et al (58) at the Franco-German
Research Institute and, currently in progress, are
the studies of Chernyi (59) in the Institute of Mechanics at the University of Moscow, and of Toong (60) at
M.I.T. The most interesting here is the onset of an
instability manifested by a distinct pattern of regular periodic wave oscillations that are set in behind
the bow wave, providing a most convenient means for
the observation of the non-steady gasdynamic feedback
effects associated with exothermic processes. These
studies are still in too early a stage to assess the
full importance of their potential, but they should
be certainly considered of interest as possible
sources of essential information on the mechanism of
self-sustenance.

2. 5. Recent Theoretical Advances.

Besides the availability of modern experimental facilities that enabled us to observe wave
phenomena generated by fast exothermic processes with
an adequate resolution in time and space, one of the
most important reasons why the subject of the proposed
study became of topical significance is the progress
made recently in our knowledge on the kinetics of chem—

ical reactions and on the gasdynamics of blast waves.

   With respect to the former, following a decade of extensive studies of the induction period, as exemplified by the work of Schott and Kensey (61), Skinner and Ringrose (62) and Wakefield, Ripley and Gardiner (63), a good deal of data has been obtained recently on the elementary reconstitution steps that are associated with the actual exothermic processes of the reaction. Of major significance in this connection are the investigations of Getzinger carried out in collaboration with Schott (64) and Blair (65).

   As to the latter, the fundamental equations of blast wave theory have been recently reformulated to make them particularly well suited for the analysis of non-steady flow fields generated by explosions in chemically reactive media (66). Salient features of this novel method of approach are presented here in Chapter 4.

2. 6. Summary and Conclusions.

   In contrast to the steady state that is sought in conventional shock tube experiments, blast waves provide non-steady flow fields which are associated with well defined rarefaction or compression wave regions. The exothermic reactions manifest them-

selves gasdynamically by becoming coupled with the
flow field in the form of closed-loop, feedback systems that provide means for energy exchange between
the reaction centers and the blast waves they generate.

Although research in this field has
been hardly started, a number of experimental techniques based on the use of blast waves have been developed for the study of exothermic reactions in either
the rarefaction fields, characteristic of decaying
blast waves, or in compression fields, associated
with wave amplification.

In connection with the former, use has
been made of such devices as the divergent test sections in shock or detonation tubes, employment of
proper test gases, as in marginal detonations, and a
variety of explosion systems, from finite source explosion apparatus to devices where virtually point explosions are obtained by local breakdown initiated by
means of focused laser irradiation.

Associated with the latter are detona-
tion tubes where pressure waves are generated by accelerating flames or by exothermic reactions developed
behind reflected shocks,as well as in a variety of
converging shock and implosion vessels.

Of particular importance to the study

of the mechanism of wave sustenance have been experimental facilities associated, in principle, with
steady flow, such as the supersonic jet of a reactive
mixture where the combustion wave is sustained behind
the Mach disc, and the ballistic range filled with an
explosive gas mixture where the reaction is maintained
behind the bow wave in front of a supersonic projectile.

One may hope that the knowledge gained
from such studies will enhance our understanding of the interrelationship existing between gasdynamic and chemical
kinetic processes which occur in explosion phenomena,
especially in the course of their most significant
period, that is when the reactions are actually associated with exothermic processes, and that this may
eventually lead to a direct exploitation of explosions
for the generation of useful, rather than destructive,
energy at the high power density levels attainable by
the use of explosively reacting systems.

References.

1. Taylor, Sir Geoffrey, "The Formation of a Blast
          Wave by a Very Intense Explosion" first
          published in Bristish Report RC-210,
          June 27, 1941 ; revised version in
          Proc. Roy. Soc., London, A201, Part I,
          pp. 159-174, Part II, pp. 175-186,
          March 1950.

2. Von Neumann, J., "The Point Source Solution",
          first published in NDRC, Div. B. Rept.
          AM-9, June 30, 1941 ; then in Shock
          Hydrodynamics and Blast Waves (ed. H.A.
          Bethe) AECD-2860, 1944 ; revised ver-
          sion in Blast Waves (ed. H.A. Bethe)
          Los Alamos Sci. Lab. Rep. LA-2000, 27-
          55, 1947 ; reprinted in John von Neu-
          mann Collected Works (ed. A.H. Taub),
          VI, 219-237, Pergamon Press, New York,
          1963.

3. Sedov, L.I., "Rasprostraneniye sil'nykh vzryvnykh
          voln" (Propagation of Intense Blast
          Waves), Priklednaya matematika i mekha-
          nika (Applied Mathematics and Mechanics),
          10, 2, 1946 ; revised version in Simi-
          larity and Dimensional Method in Mechan
          ics, Fourth Printing, Gostekhizdat,
          Moscow, 1957 (Transl.:M.Friedman, Ed.:
          M. Holt, Academic Press, New York, 363
          pp., 1959).

4. Stanyukovich, K.P., Neustanovivschiyesya dvizheniya
          splshnoy sredy (Unsteady Motions of Con
          tinuous Media), Gostekhizdat, Moscow,

1955 ; Transl.: J.G. Adashko, Ed.: M.
Holt, Pergamon Press, 745 pp., London
1960).

5. Korobeynikov, V.P., Mil'nikiva, N.S. and Ryazanov,
Ye. V., Teoriya Tochechnovo Vzryva (
(The Theory of Point Explosions), Mos-
cow, 1961, 332 pp. (Transl.: U.S. de-
partment of Commerce, JPSR: 14, 334,
Washington, D.C., July 1962).

6. Sakurai, Akira, "Blast Wave Theory," Basic Devel-
opments in Fluid Dynamics (Ed.: M.
Holt), I, 309-375, Academic Press,
New York 1965.

7. Zel'dovich, Ya. B. and Raizer, Yu. P., Fizika
Udarnykh Voln i Vyskotemperaturnikh
Hidrodinamicheskikh Yavlenii (Physics
of Shocks Waves and High Temperature
Hydrodynamic Phenomena), Gos. Izd.
Fiz. Mat. Literatury, 632 pp., Moscow,
1963 ; (Transl., ed. by W.D. Hayes and
R.F. Probstein, Academic Press, 916 pp.
New York, 1966-67).

8. Berthelot, M. and Vielle, P., "Sur la Vitesse de
propagation des phénomènes explosifs
dans les gaz," C.R. Acad. Sc., Paris,
94, 101-108, séance du 16 Janvier,
1882; 822-823, séance du 27 Mars,1882;
95, 151-157, séance du 24 Juillet,1882.

9. Mallard, E. and Le Chatelier, H.,"Recherches Expé-
rimentales et Théoriques sur la Com-
bustion des Mélanges Gaseux Explosifs",
Ann. Mines, 8, 4, 274-568, 1883.

10. Hirschfelder, J.O. and Curtiss, C.F.,"Theory of
Detonation, I. Irreversible Unimolec-

ular Reaction,"J. Chem. Phys.,28, 6 pp.
1130-1147, June 1958; II (with Linder,
B.), J. Chem. Phys.,28,6, 1147-1151,
June 1958; III (with Barnett, M.P.),J.
Chem. Phys.,30,2, 470-492, February
1959.

11. Wood, W.W. and Salsburg, Z.W.,"Analysis of Steady
State Supported One-Dimensional Detona
tions and Shocks,"Phys. Fluids,3, 4,
549-566, July-August, 1960; Wood, W.W.,
"Existence of Detonations for Small
Values of the Rate Parameter,"Phys.
Fluids, 4, 1, 46-60, January 1961.

12. Erpenbeck, Jerome J.,"Two-Reaction Study Detonations,"Phys. Fluids, 4, 4, 481-492,
April 1961; "Structure and Stability of
the Square-Wave Detonation,"Ninth Symposium (International) on Combustion;
442-450, The Combustion Institute, Academic Press, New York 1963.

13. Oppenheim, A.K. and Rosciszewski, J.,"Determination of the Detonation Wave Structure,"
Ninth Symposium (International) on Combustion, 424-434, The Combustion Institute, Academic Press, New York 1963.

14. White, D.R.,"Turbulent Structure of Gaseous Detonation," Phys. Fluids, 4, 465-480,
April 1961.

15. Shchelkin, K.K. and Troshin, Ya.K., Gazodinamika
Goreniya (Gasdynamics of Combustion),
Izs Akad Nauk SSSR, 255pp., Moscow,
1964 (Transl. NASA TTF-231,1963, and
Mono Book Corp., Baltimore, 1965).

16. Soloukhin, R.I., Udarnye Volny i Detonatsii v

Gazakh (Shock Waves and Detonation in Gases), Gos.Izd.Fiz.Mat. Literatury, 175 pp., Moscow, 1963; (Transl.: B.W. Kuvshinovv,Mono Book Corp., Baltimore, 1966).

17. Voitsekhovsky,B.V.,Mitrofanov, V.V. and Topchian, M.E., Struktura Fronta Detonatsii v Gazakh (Structure of Detonation Fronts in Gases), Izd.Sib.Otd.A.N.SSSR, 168 pp., Novosibirsk, 1963.

18. Strehlow, R.A.,"Gas Phase Detonations:Recent Devel600opments,"Combustion and Flame, 12, 2, 81-101, April 1968.

19. White, D.R. (With Cary, K.H.),"Structure of Gaseous Detonation,II.Generation of Laminar Detonation,"Phys.Fluids, 6, 5, 749-750, May 1963 ; III. "Density in the Induction Zone of Hydrogen Detonation," Phys. Fluids, 6, 7, 1011-1015, July 1963.

20. Lundstrom, E.A. and Oppenheim, A.K., "On the Influence of Non-stediness on the Thickness of the Detonation Wave," Proc. Roy. Soc., A 301, 463-478, 1969.

21. Glass, I.I., "Aerodynamics of Blast," Canadian Aeronautical Journal, 7, 3, 109-135, March 1961.

22. Boyer, D.W., "An Experimental Study of the Explosion Generated by a Pressurized Sphere," J. Fluid Mech., 9, Part 3, 401-429, 1960.

23. Oshima, K., "Blast Waves Produced by Exploding Wires," Exploding Wires, 2, 159-174 (Chace, W. and Moore, H., Ed.);

Plenum Press, Inc., New York, 1962.

24. Ramsden, S.A. and Savic, P., "A Radiative Detonation Model for the Development of a
Laser-Induced Spark in Air," Nature,
203, 1217-1219, September 19, 1964.

25. Lee, John H. and Knystautas, R., "Laser Spark
Ignition of Chemically Reactive Gases,"
AIAA Paper No. 68-146, AIAA 6th Aero-
space Sciences Meeting, New York,
January 22-24, 1968.

26. Back, G.G., Knystautas, R. and Lee, J.H., "Direct
Initiation of Spherical Detonations in
Gaseous Explosives," Proc. Twelfth Sym-
posium (International) on Combustion,
Poitiers, France, July 14-20, 1968.

27. Soloukhin, R.I. and Ragland, K.W., "Ignition Processes in Expanding Detonations," Combustion and Flame, 13, 3, 295-302,
June 1969.

28. Riemann, B., "Über die Fortpflanzung ebener Luft-
wellen von endlicher Schwingungsweite,"
Abhandlungen der Mathematischen Classe
der Königlichen Gesellschaft der Wissen-
schaften zu Götingen, 8, 43-65, 1859.

29. Friedrichs, K.O., "Formation and Decay of Shocks
Waves," Communications on Applied Mathematics, 1, 3, 211-245, September 1948.

30. Chu, B.T., "On the Generation of Pressure Waves at
a Plane Flame Front," Fourth Symposium
(International) on Combustion, 603-612,
The Williams & Wilkins Co., Baltimore,
1953; "Mechanism of Generation of Pressure Waves at Flame Fronts," NACA TN
3683, 20 pp., 1956.

31. Jones, H., "Accelerated Flames and Detonation in Gases", Proc. Roy. Soc., A248, 1254, 333-349, 1958.

32. Laderman, A.J. and Oppenheim, A.K., "Initial Flame Acceleration in an Explosive Gas," Proc. Roy. Soc., A268, 153-180, 1962.

33. Urtiew, P.A., Laderman, A.J. and Oppenheim, A.K., "Dynamics of the Generation of Pressure Waves by Accelerating Flames," Tenth Symposium (International) on Combustion, 797-804, The Combustion Institute, Pittsburgh, Pa 1965.

34. Voevodsky, V.V. and Soloukhin, R.I., "On the Mechanism and Explosion Limits of Hydrogen-Oxygen Chain Self-Ignition in Shock Waves," Tenth Symposium (International) on Combustion, 279-283, The Combustion Institute, Pittsburgh, Pa., 1965.

35. Strehlow, Roger A. (with Cohen, Arthur), "Initiation of Detonation", Phys. Fluids, 5, 1, 97-101, January 1962; (with Dyner, Harry B.), One-Dimensional Detonation Initiation," AIAA Journal, 1, 3, 591-595, March 1963.

36. Gilbert, R.B. and Strehlow, R.A., "Theory of Detonation Initiation Behind Reflected Shock Waves," AIAA Journal, 4, 10, 1966.

37. Barthel, H.O. and Strehlow, R.A., "Wave Propagation in One-Dimensional Reactive Flows," Phys. Fluids, 9, 10, 1966.

38. Perry, Robert W. and Kantrowitz, Arthur, "The Production and Stability of Converging

Shock Waves", J.Appl. Phys., 22, 7, 878-886, July, 1951.

39. Daiber, J.W., Hertzberg, A. and Wittliff, C.E., "Laser-Generated Implosions", Phys. Fluids, 9, 3, 617-619, March 1966.

40. Lee, J.H., Lee B.H.K. (With Shanfield, I.), "Two-Dimensional Unconfined Gaseous Detonation Waves," Tenth Symposium (international) on Combustion, 805-815, The Combustion Institute, Pittsburg, Pa., 1965, "Cylindrical Imploding Shocks Waves," Phys. Fluids, 8, 12, 2148-2152, December, 1965.

41. Lee, B.H.K., "Nonuniform Propagation of Imploding Shocks and Detonations", AIAA Journal, 5, 11, 1997-2003, November 1967.

42. Knystautas, R., Lee, B.H.K. and Lee, J.H.,"Diagnostic Experiments on Converging Detonations," Proceedings of the 6th International Shock Tube Symposium, Phys. Fluids Supplement, 12, 5, 1165-1168, May 1969.

43. Lee, J.H., "Collapsing Shock Waves in a Detonating Gas", Proceedings of the First International Colloquium on Gasdynamics of Explosions, Astronautica Acta, 14, 5, 421-425, 1969.

44. Knystautas, R. and Lee, J.H., "Experiments on the Stability of Converging Cylindrical Detonations," Presented at the 1968 Annual Meeting of the A.P.S. Division of Fluid Dynamics, to be published in Phys. Fluids.

45. Strehlow, R.A., Crooker, A.J. and Cusey, R.E.,
    "Detonation Initiation Behind an
    Accelerating Shock Wave", Combustion
    and Flame, 11, 4, 339-351, August 1967.

46. Glass, I.I., "Research Frontiers at Hyperveloci-
    ties", Canadian Aeronautics and Space
    Journal, 13, 8, 347-366 and 9, 1638-
    1645, September 1966.

47. Flagg, R.F. and Glass, I.I., "Explosive-Driven,
    Spherical Implosion Waves", Phys.
    Fluids, 11, 10, 2282-2284, 1968.

48. Busch, C.W., Laderman, A.J. and Oppenheim, A.K.,
    "Pressure Wave Generation in Particle-
    Fueled Combustion Systems: I. Para-
    metric Study", AIAA Journal, 4, 9,
    1638-1645, September 1966.

49. Busch, C.W., Warnock,A.S., Laderman, A.J. and
    Oppenheim, A.K., "Pressure Wave Gene-
    ration in Particle-Fueled Combustion
    Systems:II. Influence of Particle
    Motion, "AIAA Journal, 6, 2, 286,291,
    February 1968.

50. Smith, H.P., Jr., Busch, C.W. and Oppenheim, A.K.
    "Pressure Wave Generated in a Fission-
    able Gas by Neutron Irradiation, "
    Phys. Fluids, 7, 5, 676-683, May 1964.

51. Podney, W.N., Smith, Harold P., Jr., and Oppenheim,
    A.K., "On the Generation of Pressure
    Waves in Fissioning Gases", Proceed-
    ings, XXXVI International Congress of
    Industrial Chemistry,Brussels, 1966 ;
    Gr. IV, S. 12-406,1968.

52. Podney, W.N., Smith, H.P., Jr., and Oppenheim, A.K.
    "On the Generation of a Fissioning

Plasma in a Shock Tube", Proceedings
of the 6th International Shock Tube
Symposium, Phys. Fluids Supplement, 12,
5. 168-172, May 1969.

53. Zajac, L.J. and Oppenheim, A.K., "Dynamics of an
Explosive Reaction Center," Combustion
and Flame (in press).

54. Gross, R.A., "Research on Supersonic Combustion",
ARS Journal, 29, 1, 63-64, January 1959.

55. Nicholls, J.A., "Standing Detonation Waves", Ninth
Symposium (International) on Combustion,
488-496, The Combustion Institute, Academic Press, New York 1963.

56. Richmond, J. Kenneth and Shreeve, Raymond P.,
"Wind-Tunnel Measurements of Ignition
Delay Using Shock-Induced Combustion",
AIAA Journal, 5, 10, 1777-1784, October
1967.

57. Ruegg, F.W. and Dorsey, W.W., "A Missile Technique for the Study of Detonation Waves",
J. Research, Natl. Bureau of Standards,
66C, 51-58, 1962.

58. Behrens, H., Struth, W. and Wecken, F., "Studies
of Hypervelocity Firings into Mixtures
of Hydrogen with Air or with Oxygen",
Tenth Symposium (International) on Combustion, 245-252, The Combustion Institute, Pittsburgh, Pa., 1965.

59. Chernyi, G.G., "Supersonic Flow Around Bodies with
Detonation and Deflagration Fronts,"
Astronautica Acta, 13, 5 and 6, 467-480,
August 1968.

60. McVey, J.B. and Toong, T.Y., "Mechanism of Insta-

bilities of Exothermic Hypersonic Blunt-Flows", (in preparation for publication; so far a available only in the D. Sc. Thesis of McVey).

61. Schott, G.L. and Kinsey, J.L., "Kinetic Studies of Hydroxyl Radicals in Shock Waves, II. Induction Times in the Hydrogen-Oxygen Reaction", J. Chem. Phys., 29, 1179-1182, 1958.

62. Skinner, G.B. and Ringrose, G.H., "Ignition Delays of a Hydrogen-Oxygen-Argon Mixture at Relatively Low Temperatures", J. Chem. Phys., 42, 2190-2192, 1965.

63. Wakefield, C.B., Ripley, D.L. and Gardiner, W.C., "Chemical Kinetics of the Shock-Initiated Combustion of Hydrogen at High Pressure and Low Temperature", J. Chem. Phys., 50, 325-332, 1969.

64. Getzinger, R.W. and Schott, G.L., "Recombination via the H + O + M HO + M Reaction in Lean Hydrogen-Oxygen Mixtures", J. Chem. Phys., 43, 1965.

65. Getzinger, R.W. and Blair, L.S., "Recombination in the Hydrogen-Oxygen Reaction: A Shock Tube Study with Nitrogen and Water Vapor as Third Bodies", Combustion and Flame, 13, 3, 271-284, 1969.

66. Oppenheim, A.K., Lundstrom, E.A. and Kamel, M.M. "A Systematic Exposition of the Conservation Equations for Blast Waves", to be published in the Journal of Applied Mechanics.

Chapter 3.
Gasdynamic Discontinuities.

The Most Prominent Effects of Explosions.

3. 1. Introduction.

The most significant feature of flow fields generated by explosions is the existence of gasdynamic discontinuities. In the analysis they appear in the form of shocks waves, detonations and deflagrations, as well as some simple waves, especially rarefactions, that is continuous flow regions whose extent in the physical time-space domain is small enough, in comparison to the whole flow field under consideration, so that they can be treated effectively as discontinuities.

For proper interpretation of experimental records-the primary objective of the theory presented here-it is of special importance to take account of the thermodynamic properties of the medium as realistically as possible. Thus, in contrast to the classical treatment, as presented for instance in Refs. 1 - 6, all the basic relationships are introduced here without any restrictions associated with a

particular form of the equation of state, let alone
the usual idealization of a perfect gas with constant
specific heats.

Presented in this chapter are both the
static and dynamic properties of discontinuities, the
former referring to those associated with their constant velocity of propagation, and the latter to the
processes associated with velocity changes. As they
are most usually encountered in experimental observations,the dynamic events are usually instantaneous in
comparison to the periods of time during which the
discontinuitirs are "static" that is when the change
of state across them is not affected by the variation
in their propagation velocity. The consideration of
the dynamic processes is thus restricted here to the
treatment of wave interactions.

Included here also is the theory of
oblique discontinuities and of the processes of wave
intersections, the effects equivalent to interactions
that occur in a two-dimensional steady flow. This is
culminated by the analysis of interactions between
wave intersections, notably collisions between wave
intersection points, the processes which, as manifested by Fig. 1.5, play the most important role in the
dynamic behavior and structure of detonation waves.

3. 2. Mechanical Conditions.

A gasdynamic discontinuity appears in the flow field as a plane across which a finite change of state occurs without any variation in the mass flow rate per unit area, $\dot{m}$, nor in the stream force per unit area, f, that is :

$$\dot{m} = \rho_i v_i = \rho_j v_j \qquad (3.1)$$

and

$$f = p_i + \dot{m} v_i = p_j + \dot{m} v_j \qquad (3.2)$$

In the above subscripts i and j denote conditions immediately ahead and behind the discontinuity, p and ρ have the usual meaning of pressure and density, while v is the normal component of the relative flow velocity. The velocity changes occurring across a discontinuity are depicted on Fig. 3.1.(see page 101) for both the cases of a spatially one-dimensional, nonsteady flow and of a two-dimensional, steady flow field. The absolute wave velocities are denoted there by symbol w, while the particle velocities by u. As demonstrated by Fig. 3.1.b, the change in the latter is related to that of a relative flow velocity, so that, in view of Eq. (3.1.), one has :

(3.3)
$$\Delta u = u_j - u_i = v_i - v_j = v_i \left(1 - \frac{\rho_i}{\rho_j}\right)$$

while, at the same time, Eq. (3.2) yields

(3.4)
$$\frac{p_j}{p_i} = 1 + \frac{\dot{m}}{p_i} \Delta u \quad .$$

In order to express these equations in the usual non dimensional form without invoking the restrictions associated with a particular form of the equation of state, special care has to be given to the introduction of the velocity of sound. This is accomplished here by the use of a properly interpreted isentropic index, treated for this purpose as a velocity of modulus, namely :

(3.5)
$$\Gamma \equiv \left(\frac{\partial \ln p}{\partial \ln \rho}\right)_s = \left(\frac{\partial h}{\partial e}\right)_s = \frac{a^2}{p/\rho}$$

where s, h and e denote, respectively, the entropy, enthalpy and internal energy.

Besides the straightforward expressions for the pressure and specific volume ratios

$$P \equiv \frac{p_j}{p_i} \qquad \text{and} \qquad \gamma \equiv \frac{\rho_i}{\rho_j}$$

the normal Mach number and particle velocity have
thus the following meaning

$$M_n \equiv \frac{v_i}{a_i} = \frac{1}{\sqrt{\Gamma_i}} \frac{v_i}{\sqrt{P_i/\varrho_i}} \qquad (3.6)$$

and

$$U_i \equiv \frac{\Delta u}{a_i} = \frac{1}{\sqrt{\Gamma_i}} \frac{\Delta u}{\sqrt{P_i/\varrho_i}} \qquad . \qquad (3.7)$$

In terms of the above, Eq. (3.3) becomes

$$U_i = M_n(1-\nu) \qquad (3.8)$$

while Eq. (3.4) yields

$$P = 1 + \Gamma_i U_i M_n \qquad (3.9)$$

whence

$$\Gamma_i M_n^2 = \frac{P-1}{1-\nu} \qquad (3.10)$$

and

$$\Gamma_i U_i^2 = (P-1)(1-\nu) \qquad . \qquad (3.11)$$

Finally, the non-dimensional local

velocity of sound is expressed simply in terms of

$$(3.12) \qquad A_i = \frac{a_j}{a_i} = \sqrt{\frac{\Gamma_j}{\Gamma_i} P \gamma} \ .$$

3.3. The Hugoniot Curve.

The energy equation for a discontinuity can be, in general, expressed as follows :

$$(3.13) \qquad h_j - h_i - q = \frac{1}{2}(v_i^2 - v_j^2)$$

where $h_j - h_i$ represents that component in enthalpy increase which can be measured by the change in the pressure-density ratio, while q represents the rest. Introducing then

$$(3.14) \qquad H = \frac{h_j}{p_i/\rho_i} \qquad \text{while} \qquad H_i = \frac{h_i}{p_i/\rho_i}$$

and

$$(3.15) \qquad Q \equiv \frac{q}{p_i/\rho_i}$$

Equation (3.13), with the help of Eqs. (3.10) and (3.11), becomes

$$(3.16) \qquad H - (H_i + Q) = \frac{1}{2}(P - 1)(\gamma + 1).$$

The above represents the well known equation of the Hugoniot curve, expressed in a non-

dimensional form. Examples of its plots on the pressure-specific volume plane are given by Fig. 3.2.(see
page 101). On the diagram H represents the Hugoniot
curve, while RH is the corresponding Rankine-Hugoniot
curve, that is one of the same family as the other
passing through the point representing the initial
state.Points J and K represent the well-known Chapman-
Jouguet states for detonation and deflagration, respec
tively, and N-the so called von Neuman spike. The
straight lines from point (1,1) to J and to K are the
Rayleigh lines representing, in effect, plots of Eq.
(3.10) for M_n = const.

3. 4. The Chapman - Jouguet Condition

The Chapman-Jouguet condition implies
that, when the Hugoniot curve is tangent to the
Rayleigh line passing through the point representing
the initial state, it is also tangent to the isentrope.
Hence, with reference to Fig. 3.2, the local Mach
number at state J or K is unity irrespective of the
thermodynamic properties of the medium. The general
validity of this property is demonstrated here on the
basis of the classical arguments of Jouguet (7) and
Becker (8).

The proof is based on the postulate

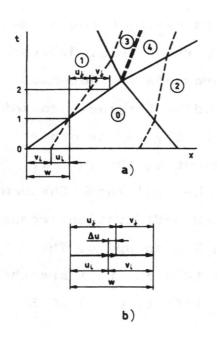

a)

b)

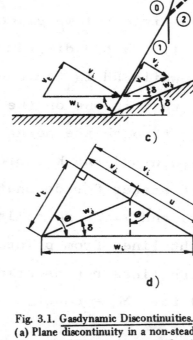

c)

d)

Fig. 3.1. Gasdynamic Discontinuities.
(a) Plane discontinuity in a non-steady, one-dimensional flow, and an example of wave interaction.
(b) Velocity diagram for plane discontinuity.
(c) Oblique discontinuity in a steady, two-dimensional flow, and an example of wave intersection.
(d) Velocity holograph for oblique discontinuity.

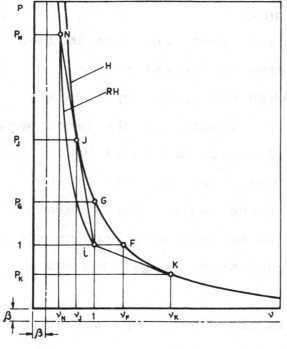

Fig. 3.2. The Hugoniot Curve on a Non-Dimensional Pressure-Specific Volume Plane.

that, in the vicinity of states J and K, the Hugoniot
curve is a locus of $Q = $ const. On the basis of Eq.(3.16),
the differentials along this curve are therefore related as follows :

$$\left(\frac{\partial H}{\partial P}\right)_\nu \left(\frac{\partial P}{\partial \nu}\right)_H + \left(\frac{\partial H}{\partial \nu}\right)_P = \frac{\nu + 1}{2} \left(\frac{\partial P}{\partial \nu}\right)_H + \frac{P - 1}{2} \qquad (3.17)$$

Hence, the slope of the Hugoniot curve
on the $P - \nu$ plane is :

$$\Phi_H \equiv -\left(\frac{\partial P}{\partial \nu}\right)_H = \frac{\left(\frac{\partial H}{\partial \nu}\right)_P - \dfrac{P-1}{2}}{\left(\frac{\partial H}{\partial P}\right)_\nu - \dfrac{\nu+1}{2}} \qquad (3.18)$$

The slope of the Rayleigh line passing
through the point representing the initial state is,
as a consequence of Eq. (3.10),

$$\Phi_R \equiv -\left(\frac{\partial P}{\partial \nu}\right)_R = \frac{P-1}{1-\nu} \qquad (3.19)$$

The Second Law, on the other hand,
demands that for an isentropic process

$$\left(\frac{\partial H}{\partial P}\right)_\nu dP + \left(\frac{\partial H}{\partial \nu}\right)_P d\nu - \nu\, dP = 0 \qquad (3.20)$$

so that the slope of the isentrope is :

$$(3.21) \qquad \Phi_S \equiv -\left(\frac{\partial P}{\partial v}\right)_S = \frac{\left(\frac{\partial H}{\partial v}\right)_P}{\left(\frac{\partial H}{\partial P}\right)_v - v} \; .$$

As the point of tangency between the Hugoniot curve and the Rayleigh line

$$(3.22) \qquad \Phi_H = \Phi_R$$

or, by virtue of Eqs. (3.18) and (3.19)

$$\left[\left(\frac{\partial H}{\partial P}\right)_v - \frac{v+1}{2}\right](P-1) = \left(\frac{\partial H}{\partial v}\right)_P (1-v) - \frac{1-v}{2}(P-1)$$

or

$$(3.23) \qquad \frac{P-1}{1-v} = \frac{\left(\frac{\partial H}{\partial v}\right)_P}{\left(\frac{\partial H}{\partial P}\right)_v - v} \; .$$

The above, according to Eqs. (3.19) and (3.21), means that

$$(3.24) \qquad \Phi_R = \Phi_S$$

which, in view of Eq. (3.22), completes the proof of the Chapman-Jouguet condition.

3. 5. Parametric Forms of the Hugoniot Equations.

The Hugoniot equation can be expressed in a parametric form by the introduction of a parameter β in terms of which

$$H \equiv \frac{1+\beta}{2\beta} P\nu \qquad \text{while} \qquad H_i \equiv \frac{1+\beta_i}{2\beta_i} . \qquad (3.25)$$

Thus, Eq.(3.16) becomes simply

$$(P + \beta)(\nu - \beta) = C \qquad (3.26)$$

where, with reference to Fig. 3.2,

$$C = (1 + \beta)(\nu_F - \beta)$$

or

$$C = (1 - \beta)(P_G + \beta)$$

so that

$$\nu_F = \frac{2\beta}{1+\beta} \left(\frac{1+\beta_i}{2\beta_i} + Q \right) \qquad (3.27)$$

is the specific volume or velocity ratio associated with a constant pressure deflagration, while

$$P_G = \frac{2\beta}{1-\beta} \left(\frac{1-\beta_i}{2\beta_i} + Q \right) \qquad (3.28)$$

is the pressure ratio corresponding to combustion at constant volume.

If β = const. and C = const, Eq. (3.26) represents a rectangular hyperbola, as demonstrated on Fig. 3.2.

This notion is of central importance to a proper interpretation of the Hugoniot relation. It points out that the most appropriate manner of interpolating between points on the Hugoniot curve is by a hyperbolic fit. Moreover, it provides an opportunity for a great simplification of the computational procedure if a single hyperbola can cover a significant range of operating conditions which, in our experience, has proven indeed to be the case.

The accuracy of the fit can be improved by noticing that Eq. (3.18) can also be written as

$$(3.29) \qquad (P + \beta_P)(v - \beta_v) = C$$

where $\beta_P \neq \beta_v$ while, similarly as before $C =$
$= (1 + \beta_P)(v_F - \beta_v) = (1 - \beta_v)(P_G + \beta_P) = $ const.

Thus, for a three-point fit, by eliminating C from the three equations of the form of Eq. (3.29) for P_n and v_n ($n = i, j, k$) one obtains :

$$(3.30a) \qquad \beta_P = \frac{(P_k - P_j)P_i v_i + (P_i - P_k)P_j v_j + (P_j - P_i)P_k v_k}{(v_k - v_j)P_i + (v_i - v_k)P_j + (v_j - v_i)P_k}$$

and

$$\beta_\nu = \frac{(\nu_k - \nu_j)P_i\nu_i + (\nu_i - \nu_k)P_j\nu_j + (\nu_j - \nu_i)P_k\nu_k}{(\nu_k - \nu_j)P_i + (\nu_i - \nu_k)P_j + (\nu_j - \nu_i)P_k} \qquad (3.30b)$$

For the Rankine-Hugoniot Curve $P_i = \nu_i = 1$

whence

$$\beta_P = \frac{(P_k\nu_k - 1)(P_j - 1) - (P_j\nu_j - 1)(P_k - 1)}{(1 - \nu_k)(P_j - 1) - (1 - \nu_j)(P_k - 1)} \qquad (3.31a)$$

and

$$\beta_\nu = \frac{(1 - \nu_k)(P_j\nu_j - 1) - (1 - \nu_j)(P_k\nu_k - 1)}{(1 - \nu_k)(P_j - 1) - (1 - \nu_j)(P_k - 1)} \qquad (3.31b)$$

Moreover, from Eq. (3.29) it follows
that, for any point n on the Hugoniot hyperbola,

$$\Phi_H \equiv -\left(\frac{\partial P}{\partial \nu}\right)_H = \frac{P_n + \beta_P}{\nu_n - \beta_\nu} \qquad (3.32)$$

while, for any point j on the Rayleigh line one has,
on the basis of Eq. (3.10)

$$\Phi_R \equiv -\left(\frac{\partial P}{\partial \nu}\right)_R = \frac{P_j - 1}{1 - \nu_j} = \Gamma_j \frac{P_j}{\nu_j} M_j^2 = \Gamma_i M_n^2 \qquad (3.33)$$

Thus, for the Chapman-Jouguet state, J (or K), at
which the Rayleigh line is tangent to the Hugoniot

curve, while the local Mach number is unity, one obtains :

$$(3.34) \qquad \Gamma_j = \frac{1 + \beta_v}{1 - \beta_v} + \frac{\beta_P - \beta_v}{1 - \beta_v} \frac{1}{P_J} \; .$$

For the Rankine-Hugoniot curve the Mach number is unity at state $P_i = v_i = 1$, hence from Eq. (3.32) and (3.33) it follows that

$$(3.35) \qquad \Gamma_i = \frac{1 + \beta_P}{1 - \beta_v} \; .$$

If $\beta_P = \beta_v$, or if $P_J = 1$, Eq. (3.34) becomes, of course, identical to Eq. (3.35).

At the same time, it is of interest to note that for $\Phi_H = \Phi_R$, when $P_n = P_j = P_J$ and $v_n = v_j = v_J$, Eqs. (3.32) and (3.33) yield the relation

$$(3.36) \qquad (2P_J - 1 + \beta_P)(2v_J - 1 - \beta_v) = (1 + \beta_P)(1 - \beta_v)$$

which with respect to the coordinates

$$(3.37a) \qquad P_N = 1 + 2(P_J - 1)$$

and

$$(3.37b) \qquad v_N = 1 - 2(1 - v_J)$$

is of the form of Eq. (3.29) with $P_G = v_F = 1$. Thus Eqs. (3.37) represent simple relationships between

the coordinates of the von Neumann spike and those
of the Chapman-Jouguet point. This property has been
pointed out by Langweiler(9) with reference to the
case of a perfect gas with constant specific heats.
As it appears from the above, this is just the consequence of the geometric properties of the hyperbolic
approximation to the Hugoniot curve, the doubling of
the differences between the corresponding coordinates
implied by Eqs.(3.37) being valid only if the Rankine
Hugoniot curve for the shock and the Hugoniot curve
for the detonation belong to the same family of
curves expressed in terms of Eq. (3.29).

   In the case when $\beta_P = \beta_v = \beta$, Eqs.
(3.30) are reduced to

$$\beta = \frac{(P_j - 1) v_j - (P_k - 1) v_k}{(P_j - v_j) - (P_k - v_k)} \, . \qquad (3.38)$$

For a Rankine-Hugoniot curve, this becomes reduced
further to

$$\beta = \frac{(P_j - 1) v_j}{P_j - v_j} \qquad (3.39)$$

while, with respect to the coordinates of the Chap-
mann-Jouguet state, Eq. (3.36) yields :

$$\beta = 1 - \frac{2 P_J (1 - v_J)}{P_J - v_J} \, . \qquad (3.40)$$

In the above ν_J can be expressed in terms of M_J and P_J by the use of the Rayleigh relations, Eq. (3.33), with $M_n = M_J$, $P = P_J$ and $\nu = \nu_J$, while $M_j = 1$. One obtains then :

$$(3.41a) \qquad\qquad \beta = 1 - \frac{2\,P_J}{\Gamma_i\,M_J^2 + 1}$$

or, by virtue of Eq. (3.35),

$$(3.41b) \qquad\qquad P_J = \frac{1-\beta}{2} + \frac{1+\beta}{2}\,M_J^2 .$$

Finally, it should be reiterated that, as it was indeed pointed out at the outset, all the expressions derived here for state J are also valid for state K except, of course, for Eqs. (3.37), which, with reference to the latter, lose their physical meaning.

3. 6. The Change of State Across a Discontinuity.

The parameters describing the change of state across a discontinuity are : the pressure ratio, P , the relative velocity, or specific volume ratio, ν , the front Mach number, M_n , the change in the normal particle velocity expressed in terms of U and the ratio of the local velocities of sounds, A .

The most straightforward expressions are obtained from the Hugoniot relations in terms of the pressure ratio. Thus from Eq. (3.29) one obtains

$$
\left\{
\begin{array}{l}
\nu = \beta_\nu + (1 - \beta_\nu)\dfrac{P_G + \beta_P}{P + \beta_P} = \dfrac{(P_G + \beta_P) + \beta_\nu(P + P_G)}{P + \beta_P} \\[6mm]
\\
= 1 - (1 - \beta_\nu)\dfrac{P - P_G}{P - \beta_P} \; .
\end{array}
\right.
\tag{3.42}
$$

Equations (3.10) and (3.11), combined with Eq. (3.42), yield then, respectively,

$$
M_n^2 = \frac{(P + \beta_P)(P - 1)}{(1 - \beta_\nu)(P - P_G)}
\tag{3.43}
$$

and

$$
U_i^2 = \frac{(1 - \beta_\nu)^2 (P - P_G)(P - 1)}{(1 + \beta_P)(P + \beta_P)}
\tag{3.44}
$$

while, according to Eq. (3.12),

$$
A_i^2 = \frac{(P_G + \beta_P) + \beta_\nu(P - P_G)}{(1 + \beta_P)(P + \beta_P)}(1 - \beta_\nu)\Gamma P .
\tag{3.45}
$$

In most applications the change of state across the discontinuity has to be evaluated in terms of its Mach number. For this purpose one gets from Eq. (3.43)

$$(3.46) \qquad P = \frac{(1 + \beta_P) M_n^2 + (1 - \beta_P)}{2} \pm \Delta$$

where

$$(3.47) \qquad \Delta^2 = \left[\frac{(1 + \beta_P) M_n^2 + (1 - \beta_P)}{2} \right]^2 - (1 + \beta_P) P_G M_n^2 + \beta_P .$$

The other parameters can be determined then by the substitution of the above expression for P in Eqs. (3.42), (3.44) and (3.45). The plus and minus signs correspond, respectively, to the upper and lower intersection between the Hugoniot curve and a Rayleigh line passing through the point representing the initial state.

At the Chapman-Jouguet state $\Delta = 0$ and Eq. (3.47) yields the following expression for the wave Mach number :

$$(3.48) \qquad M_{J,K}^2 = \frac{2 P_G - (1 - \beta_P)}{1 + \beta_P} \pm \frac{2}{1 + \beta_P} \sqrt{P_G^2 - (1 + \beta_P) P_G - \beta_P} .$$

For the Rankine-Hugoniot curve $P_G = 1$. Equations (3.43), (3.44) and (3.45) become then

$$(3.49) \qquad M_n^2 = \frac{P + \beta_P}{1 + \beta_P}$$

$$U_i^2 = \frac{(1 - \beta_v)^2 (P - 1)^2}{(1 + \beta_P)(P + \beta_P)} \qquad (3.50)$$

and

$$A_i^2 = \frac{1 - \beta_v}{1 + \beta_P} \cdot \frac{1 + \beta_P + \beta_v(P - 1)}{P + \beta_P} \Gamma P \qquad (3.51)$$

while, as it is evident directly from Eq. (3.49), Eq. (3.46) reduces to

$$P = (1 + \beta_P) M_n^2 - \beta_P . \qquad (3.52)$$

With the use of the above Eqs. (3.42), (3.50) and (3.51), yield, respectively,

$$v = \beta_v + \frac{1 - \beta_v}{M_n^2} \qquad (3.53)$$

$$U_i = (1 - \beta_v)\left(1 - \frac{1}{M_n^2}\right) M_n \qquad (3.54)$$

and

$$A_i^2 = (1 - \beta_v)^2 \left(M_n^2 - \frac{\beta_P}{1 + \beta_P}\right)\left(\frac{1}{M_n^2} + \frac{\beta_v}{1 - \beta_v}\right)\Gamma . \qquad (3.55)$$

For a perfect gas with constant specific heats $\Gamma = \Gamma_i = \gamma$ and $\beta_P = \beta_v = \dfrac{\gamma - 1}{\gamma + 1}$. Under such circumstances Eqs. (3.52) - (3.55) acquire the well known forms

$$(3.56) \qquad P = \frac{2\gamma}{\gamma+1} M_n^2 - \frac{\gamma-1}{\gamma+1}$$

$$(3.57) \qquad \nu = \frac{\gamma-1}{\gamma+1} + \frac{2}{\gamma+1} \frac{1}{M_n^2}$$

$$(3.58) \qquad U_i = \frac{2}{\gamma+1} \left(1 - \frac{1}{M_n^2}\right) M_n$$

and

$$(3.59) \qquad A_i^2 = \left(\frac{2}{\gamma+1}\right)^2 \left(\gamma M_n^2 - \frac{\gamma-1}{2}\right)\left(\frac{1}{M_n^2} + \frac{\gamma-1}{2}\right).$$

3. 7. Oblique Discontinuity.

In order to treat oblique discontinuities, one has to consider only the additional features associated solely with the flow geometry, that is with kinematic relations which are, in effect, expressed by the velocity diagram on Fig. 3.1. Since a plane discontinuity can affect only the normal component of flow, velocity changes can be represented by a hodograph where, as shown on Fig. 3.1c, the velocity vectors w_i and w_j have a common tangential component v_t.

On the basis of this diagram it follows then that the angle of incidence

$$\theta = \sin^{-1}\left(\frac{v_i}{w_i}\right) \tag{3.60}$$

the flow deflection angle

$$\delta = \cot^{-1}\left[\left(\frac{w_i}{u\sin\theta} - 1\right)\tan\theta\right] \tag{3.61}$$

and the flow velocity immediately behind the discontinuity

$$w_\delta^2 = w_i^2 + u^2 - 2v_i u \ . \tag{3.62}$$

Noting that $\frac{v_i}{a_i} = M_n$ while $\frac{w_i}{a_i} = M_i$,
Eq. (3.60) can be combined with Eq. (3.10) to yield

$$\theta = \sin^{-1}\sqrt{\frac{P-1}{\Gamma_i M_i^2 (1-\gamma)}} \ . \tag{3.63}$$

At the same time, by virtue of Eqs. (3.8) and (3.10),
Eq. (3.61) becomes

$$\delta = \cot^{-1}\left[\left(\frac{\Gamma_i M_i^2}{P-1} - 1\right)\tan\theta\right] \tag{3.64}$$

while, with the use of Eqs. (3.9) and (3.11), Eq. (3.62) gives

(3.65)
$$M_j^2 = \frac{\Gamma_i M_i^2 - (P-1)(\nu+1)}{\Gamma_j P \nu}$$

where $M_j \equiv w_j / a_j$.

In the above equations ν can be eliminated by means of the Hugoniot relation. Thus, using Eq. (3.29) with $C = (1 - \beta_\nu)(P_G + \beta_P)$ one obtains, respectively, from Eqs. (3.64), (3.63) and (3.65),

(3.66)
$$\tan^2\delta = \frac{(1+\beta_P)M_i^2(P-P_G) - (P+\beta_P)(P-1)}{[(1+\beta_P)M_i^2 - (1-\beta_\nu)(P-1)]^2} \cdot \frac{(P-1)}{P+\beta_P}(1+\beta_P)^2$$

(3.67)
$$\tan^2\theta = \frac{(P+\beta_P)(P-1)}{(1+\beta_P)M_i^2(P-P_G) - (P+\beta_P)(P-1)}$$

and

(3.68)
$$M_j^2 = \frac{(1+\beta_P)M_i^2(P+\beta_P) - (1-\beta_\nu)[(1+\beta_\nu)P + (1-\beta_\nu)P_G + 2\beta](P-1)}{\Gamma_j(1-\beta_\nu)[(P_G+\beta_P) + \beta_\nu(P-P_G)]P}$$

Whence, for the Rankine-Hugoniot relation, for which $P_G = 1$,

$$\tan^2\delta = \frac{(1+\beta_P)M_i^2 - (P+\beta_P)}{[(1+\beta_P)M_i^2 - (1-\beta_\nu)(P-1)]^2} \cdot \frac{(P-1)^2}{P+\beta_P}(1-\beta_\nu)^2$$

(3.69)

$$\tan^2\Theta = \frac{P+\beta_P}{(1+\beta_P)M_i^2 - (P+\beta_P)} \tag{3.70}$$

and

$$M_j^2 = \frac{(1+\beta_P)M_i^2(P+\beta_P)-(1-\beta_v)\left[(1+\beta_v)P+1-\beta_v+2\beta_P\right](P-1)}{\Gamma_j(1-\beta_v)\left[1+\beta_P+\beta_v(P-1)\right]P}. \tag{3.71}$$

3. 8. Simple Wave.

The salient property of a simple wave in a gasdynamic flow field is the fact that its characteristics are straight lines, while the change of state across it is isentropic. As a consequence of the latter, the equation of motion with respect to a characteristic is

$$\left(\frac{\partial u}{\partial p}\right)_s = \mp \frac{1}{a\,\rho}. \tag{3.72}$$

By the introduction of a parameter

$$\alpha = \frac{2}{\rho}\left(\frac{\partial p}{\partial a^2}\right)_s = 2\left(\frac{\partial h}{\partial a^2}\right)_s \tag{3.73}$$

the above is reduced to

(3.74)
$$\left(\frac{\partial u}{\partial a}\right)_s = \mp \alpha \ .$$

Substituting into Eq. (3.73) the expression for a given in terms of Γ by Eq. (3.5) one obtains

(3.75)
$$\frac{1}{\alpha} = \frac{p}{2}\left(\frac{\partial \Gamma}{\partial p}\right)_s + \frac{\Gamma - 1}{2} \ .$$

From the above it appears that if Γ = const.

(3.76)
$$\alpha = \frac{2}{\Gamma - 1}$$

but when α = const. Γ may be a variable. In this case Eq. (3.75) can be integrated yielding

(3.77)
$$\Gamma = \frac{\Gamma_i}{P}\left[\frac{2 + \alpha}{\alpha \Gamma_i}(P - 1) + 1\right]$$

where, as before, $P = p/p_i$ while $\Gamma_i = \Gamma$ at $P = 1$.

The process relations between pressure and specific volume can be now obtained from the expression defining Γ , Eq. (3.5) which in non-dimensional form can be expressed as

$$\Gamma = -\frac{v}{P}\left(\frac{\partial P}{\partial v}\right)_s \tag{3.78}$$

Thus, equating the expressions for Γ given by the above two equations and integrating, one obtains

$$P = \frac{\alpha \Gamma_i}{2+\alpha}\left(v^{-\frac{2+\alpha}{\alpha}} - 1\right) + 1 \tag{3.79a}$$

or

$$v = \left[\frac{2+\alpha}{\alpha \Gamma_i}(P-1)+1\right]^{-\frac{\alpha}{2+\alpha}}. \tag{3.79b}$$

The above can be recognized as the generalized polytropic relationship used to describe the behavior of isotropic elastic solids (10).

Finally, by virtue of Eq. (3.12), Eqs. (3.77) and (3.79) yield

$$A_i = \left[\frac{2+\alpha}{\alpha \Gamma_i}(P-1)+1\right]^{\frac{1}{2+\alpha}} = v^{-\frac{1}{\alpha}}. \tag{3.80}$$

Equation (3.74) becomes, in non-dimensional form

$$U_i = \mp\alpha\,(A_i - 1) \tag{3.81}$$

while the local wave propagation velocity

(3.82) $W_i = U_i \pm A_i$.

3. 9. The Prandtl - Meyer Expansion

The Prandtl-Meyer expansion is, in essence, a simple wave in steady, two-dimensional flow. The velocity diagrams of this flow and the corresponding velocity hodograph are shown on Fig. 3.3, (see page 120). From the latter it follows that

$$b = (2w \sin^2 \frac{d\delta}{2} + dw) \cot (\theta + d\delta)$$

whence, noting that in the limit second order terms vanish identically,

(3.83) $d\delta = \dfrac{b}{w} = \dfrac{\cot \theta}{w} dw$

Since $\sin \theta = \dfrac{1}{M}$, one gets then

(3.84) $d\delta = \sqrt{M^2 - 1} \dfrac{dw}{w}$

while, noting that $d\theta = d(\sin \theta)/\sqrt{1 - \sin^2\theta}$,

(3.85) $d\theta = - \dfrac{dM}{M\sqrt{M^2 - 1}}$

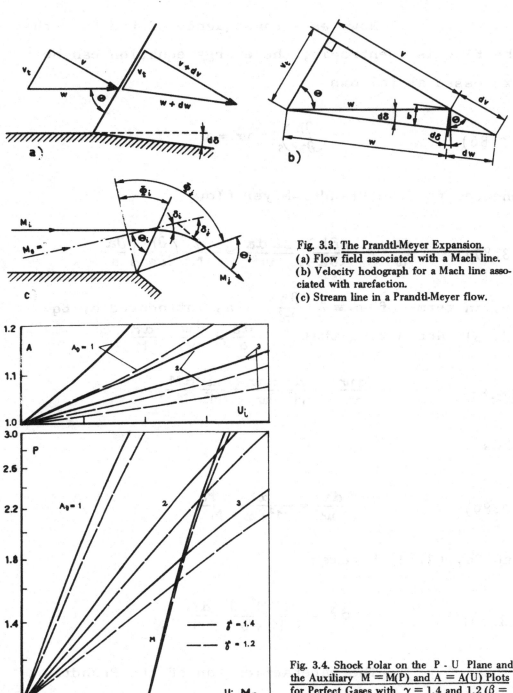

Fig. 3.3. The Prandtl-Meyer Expansion.
(a) Flow field associated with a Mach line.
(b) Velocity hodograph for a Mach line associated with rarefaction.
(c) Stream line in a Prandtl-Meyer flow.

Fig. 3.4. Shock Polar on the P - U Plane and the Auxiliary $M = M(P)$ and $A = A(U)$ Plots for Perfect Gases with $\gamma = 1.4$ and 1.2 ($\beta = 1/6$ and $1/11$).

Now, as a consequence of the fact that the flow is isentropic, the energy equation can be expressed as follows

$$(3.86) \qquad \left(\frac{\partial h}{\partial w}\right)_s + w = 0$$

whence, for the Prandtl-Meyer flow,

$$(3.87) \qquad \frac{dw}{w} = -\left(\frac{\partial h}{\partial a^2}\right)_s \frac{2a\,da}{w^2} = -\frac{2}{M^2}\left(\frac{\partial h}{\partial a^2}\right)_s \frac{\partial a}{a}$$

or, in terms of $\alpha \equiv 2\left(\frac{\partial h}{\partial a^2}\right)_s$, as introduced by Eq. (3.73), and noting that $\dfrac{da}{a} = \dfrac{dw}{w} - \dfrac{dM}{M}$,

$$(8.88) \qquad \frac{dw}{w} = -\frac{\alpha}{M^2}\frac{dw}{w} + \frac{\alpha}{M^2}\frac{dM}{M} \ .$$

Thus

$$(3.89) \qquad \frac{dw}{w} = \frac{\alpha}{M^2 + \alpha}\frac{dM}{M}$$

and Eq. (3.84) becomes

$$(3.90) \qquad d\delta = \frac{\alpha\sqrt{M^2 - 1}}{M^2 + \alpha}\frac{dM}{M} \ .$$

The rarefaction fan of the Prandtl-Meyer flow is fully described in terms of the relation ship between the angle Φ defined in Fig. 3.3d and the Mach number. According to the diagram

$$\Phi = \frac{\pi}{2} + \delta - \theta' \tag{3.91}$$

and with the use of Eqs. (3.85) and (3.90)

$$d\Phi = d\delta - d\theta = \frac{\alpha + 1}{(M^2 - 1) + (\alpha + 1)} \frac{M\,dM}{\sqrt{M^2 - 1}} \tag{3.92}$$

whence, in terms of $\mu = \sqrt{M^2 - 1}$,

$$\Phi = \int \frac{\alpha + 1}{\mu^2 - (\alpha + 1)}\, d\mu \tag{3.93}$$

If α = const., the above yields

$$\Phi = \sqrt{\alpha + 1}\, \tan^{-1} \frac{\mu}{\alpha + 1} = \sqrt{\alpha + 1}\, \tan^{-1}\sqrt{\frac{M^2 - 1}{\alpha + 1}} \tag{3.94}$$

where the constant of integration has been adjusted so that $\Phi = 0$ at $M = 1$ as indicated on Fig. 3.3d.

Thus, since $\frac{\pi}{2} - \theta = \cos^{-1}\left(\frac{1}{M}\right)$ with the use of the above one gets from Eq. (3.91)

$$\delta = \sqrt{\alpha + 1}\, \tan^{-1}\sqrt{\frac{M^2 - 1}{\alpha + 1}} - \cos^{-1}\left(\frac{1}{M}\right) \tag{3.95}$$

For a perfect gas with constant specific heats, as
pointed out by Eq. (3.76) $\alpha = \dfrac{2}{\gamma - 1}$, so that $\alpha + 1 =$
$= \dfrac{\gamma + 1}{\gamma - 1}$.

Finally, the relationship between the
local Mach number and pressure can be obtained from
the Eulers equation, which, by virtue of Eqs. (3.5)
and (3.78), can be expressed as follows

$$(3.96) \qquad \left(\frac{\partial}{\partial P} \Gamma P \gamma M^2 \right) = -\frac{2}{\gamma} .$$

For $\alpha = $ const. one can use Eqs. (3.77) and (3.79) to
eliminate Γ and γ , and Eq. (3.96), upon integration,
yields

$$(3.97) \quad M^2 = \left(\frac{2+\alpha}{\alpha \Gamma_i} \frac{P-1}{P} + \frac{1}{P} + \alpha \right) \left[\frac{2+\alpha}{\alpha \Gamma_i} (P-1) + 1 \right]^{-\frac{\alpha+1}{\alpha}} - \alpha .$$

If, moreover, $\Gamma = $ const., Eq. (3.97) reduces to the
well known form of the isentropic relation for the
perfect gas with constant specific heats, namely

$$(3.98) \qquad M^2 = \frac{\Gamma + 1}{\Gamma - 1} P^{-\frac{\Gamma-1}{\Gamma}} - \frac{2}{\Gamma - 1} .$$

3. 10. Wave Interactions.

Wave interactions are understood to occur normally in unsteady flow fields when waves are plane and their fronts are parallel to each other. Under such circumstances they can undergo either head-on or rear-end collisions. Usually the duration of such collisions is negligibly small in comparison to the rest of the flow phenomena under observation so that the results of a collision process can be evaluated by determining a stationary wave system which is dynamically compatible with the incident waves.

A collision between two wave fronts is illustrated in Fig. 3.1a. Each of the colliding partners generates a reflected wave. The thermodynamic states generated by this process are, as a rule, different on the two sides of the collision point due to the different history of particles that have been affected by the incident and reflected waves on each side. The condition of dynamic compatibility demands that the pressures and particle velocities in the two regimes (regions (3) and (4) in Fig. 3.1a) be the same. This is then the key to the solution.

For this purpose waves are considered
to act as gasdynamic discontinuities and the loci of
states attainable by them are expressed in terms of
the so-called polars, that is curves on the pressure-
flow velocity plane. The solution of an interaction
process is then obtained by finding the intersection
between proper polars. In general this yields data for
the reflected waves in terms of the states generated
by their action (e.g. states (3) and (4) in Fig. 3.1a)
and, in particular, the direction of the contact discontinuity, that is the boundary between these states
that travels at their common particle velocity.

The most convenient manner of determining such a solution is, of course, by means of a
graphical technique, especially the Vector Polar
Method (11) where the polar diagram is rendered a vec-
torial character by the use of appropriate scales for
the coordinates. The solution thus obtained on the
planes of such coordinates represents, in effect, a
vector determined by the summation of component vectors, each expressing the change of state brought
about by the action of one of the wave fronts resulting from the interaction process. Since pressure
ratios are multiplied in crossing a discontinuity
while the particle velocities are added, this is sim-

ply achieved by using a logarithmic scale for P and a linear scale for U.

Such shock polars for the case of perfect gases with $\gamma = 1.4$ and $\gamma = 1.2$ are presented on Fig. 3.4 (see page 127) together with the auxiliary diagrams of $M_n = M(P)$ and $A = A(U)$. They represent, in effect, plots of Eqs. (3.50), (3.49) and (3.51) with $\beta_p = \beta_v = \frac{1}{6}$ and $\frac{1}{11}$. Since for vector addition on the polar diagram, $U = \frac{w}{a_o}$ (where a_o is the velocity of sound in the undisturbed medium—a universal constant for all the waves in a given problem), rather than $U_i = \frac{w}{a_i}$ (where a_i is the local velocity of sound immediately ahead of a given discontinuity), must be taken into account in determining the state resulting from a given interaction process, the polars used for this purpose must be expressed in terms of

$$U = U_i A_o \qquad (3.99)$$

where $A_o = \frac{a_i}{a_o}$. The implication of this rule is quite simple : the abscissa of the appropriate polar is obtained by the multiplication specified by Eq. (3.99), as reflected by curves corresponding to $A_o = 2$ and 3 on Fig. 3.4.

The corresponding rarefaction polars

are given by Fig. 3.5 (see page 129) together with
the auxiliary $A_i = A(P)$ diagram. These are, essentially, plots of Eqs. (3.80) and (3.81). The rule of Eq.
(3.99) applies there as well, as manifested by the
polars corresponding to $A_0 = 2$ and 3.

An example of deflagration polars and
their auxiliary diagrams is presented by Fig. 3.6,
(see page 128). It applies to an equimolar hydrogen-
oxygen mixture for which, with a sufficiently good
approximation, $\gamma_i = 1.4$, $\gamma_j = 1.2$ and $q = 931$ cal/
gm. From this it follows that, with reference to Eq.
(3.26), $\beta = \beta_P = \beta_v = 1/11$ and $C = 5.55$ ($P_{G_0} = 6.01$; $v_{F_0} =$
$= 5.18$). Thus the curves for $A_0 = 1.0$ in Fig. 3.6
are essentially plots of Eqs. (3.43), (3.44) and (3.
(3.45), corresponding to the above values of β and P_G.

For $A_0 \neq 1$ in this case, however, beside the
rule of Eq. (3.99), one has also a correction for P_G,
namely

$$(3.100) \qquad P_G = B + \frac{P_{G_0} - B}{A_0^2}$$

where $B \equiv \frac{\gamma_i - 1}{\delta_i - 1}$. The rule of Eq. (3.100) is obtained
from the energy equation for constant volume proces-
ses at $v = 1$ satisfying the condition of $q = $ const.
for various initial states brought about solely by the

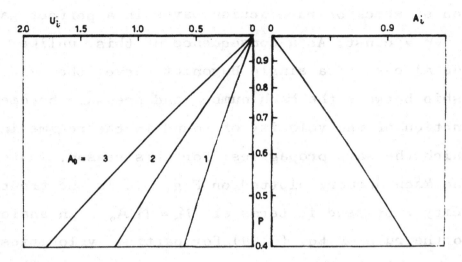

Fig. 3.5. Rarefaction Polar on the P-U Plane and the Auxiliary A = A(P) Plot for a Perfect Gas with $\gamma = 1.4\,(\alpha = 5)$.

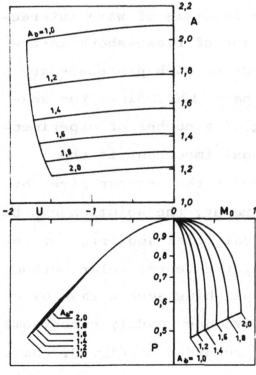

Fig. 3.6. Deflagration Polars on the P-U Plane and the Auxiliary A = A(U) and $M_o = M(P)$ Plots for $\beta = 1/11$ and C = 5.55. Corresponding to an Equimolar Hydrogen-Oxygen Mixture Initially at a Pressure of 1 atm and Room Temperature ($q = 931$ cal/gm).

action of shock or rarefaction waves in a perfect gas
with γ = const. As a consequence of this, unlike
the usual case of a single Hugoniot curve, the relationship between the Mach number and pressure becomes
a function of the velocity of sound in the regime into which the wave propagates. For this reason, instead
of the Mach number, plotted on Fig. 3.6 is the front
velocity expressed in terms of $M_o = M_n A_o$ in analo-
gy to the rule of Eq. (3.99) for particle velocities,
the value of M_n being also affected, however, by the
change in P_G implied by Eq. (3.100).

As an example of the application of the
Vector Polar Method for the analysis of wave interactions, presented here is a case of flame-shock interactions. Experimental records of such processes are
presented on Fig. 3.7 (see page 130). Since the analysis involves the matching of a number of experimental records, of which the most important is the
graphical time-space history of the process given by
the streak-schlieren photography, the solution can be
arrived at only by a graphycal trial and error procedure. In this respect, then, the Vector Polar Method
is indeed unique, for accomplishing such a task by a
numerical technique would have been unduly cumbersome.

The phenomena recorded on Fig. 3.7 have

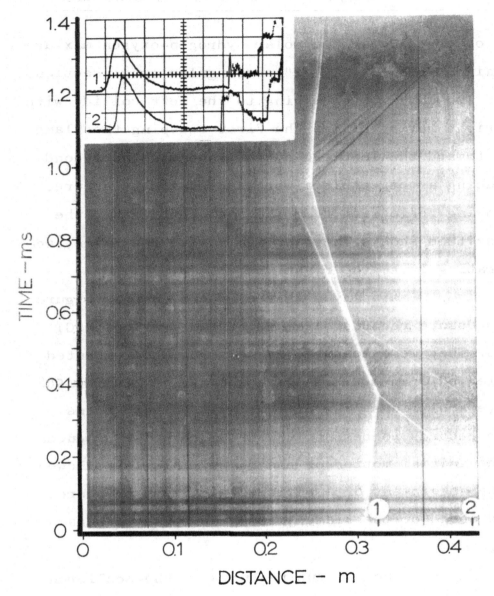

Fig. 3.7. Streak Schlieren Photograph and Pressure Transducer Records of Flame-Shock Interactions in an Equimolar Hydrogen-Oxygen Mixture Maintained Initially at a Pressure of 1 atm. and Room Temperature

Oscilloscope sweep leads the streak record by 3.38 msec; sweep rate : 0.5 msec/cm; vertical deflection ; 0.5 psi/cm.

been observed in an equimolar hydrogen-oxygen mixture
contained in a 1" x 1-1/2" rectangular cross-section
tube. Two shock waves moving to the left collide with
the right moving flame, the first sweeping the flame
back toward the left end of the tube and the second
producing a noticeable alteration of its structure.
Following reflection from the end of the tube, the
transmitted shocks interact again with the flame, accelerating it to the right.

Displayed in the insert on this figure
are pressure records obtained at two stations, PG1
and 2. They were measured by means of shock-mounted
Kistler 601 transducers with the use of Kistler 568
charge amplifiers. The first signal on both transducer records represents the pressure pulse produced
by the initial motion of the flame. This is followed
by the attainment of a relatively uniform pressure
field ahead of the flame just prior to the interaction.

The salient features of the schlieren
record of Fig. 3.7 is the appearance of relatively
straight line traces of wave fronts, providing the
justification for the finite wave interaction analysis
in which the flow is treated essentially as one-dimensional and the only changes of state admitted are

those brought about by wave action. The waves are con
sidered as plain discontinuities with the exception
of the rarefaction fan, which is treated as a zone of
continuous isentropic expansion. Thus all considera-
tions associated with the structure of the wave processes are neglected in favor of their dynamic effects.

          The solution of the wave-interaction
analysis is shown in the time-space diagram of Fig.
3.8 (see page 133) and the pressure-particle velocity
diagram of Fig. 3.9 (see page 134). On the Fig. 3.8
the flame is denoted by a double solid line and the
shock by single solid lines ; the bounds of the rarefaction fan are indicated by chain-dotted lines. For
comparison, the observations of Fig. 3.7 have been
reproduced on Fig. 3.8, the flame trace represented
by a double-dash line and the shock waves by a single-
dashed line.

          Each regime in the time-space domain
is denoted by a number which indicates its thermodynamic state on the pressure-velocity diagram. In
performing the analysis, the contact surfaces generated by each interaction were ignored, since their
effects were of second order and had negligible influence on the results.

          From experimental observations of the

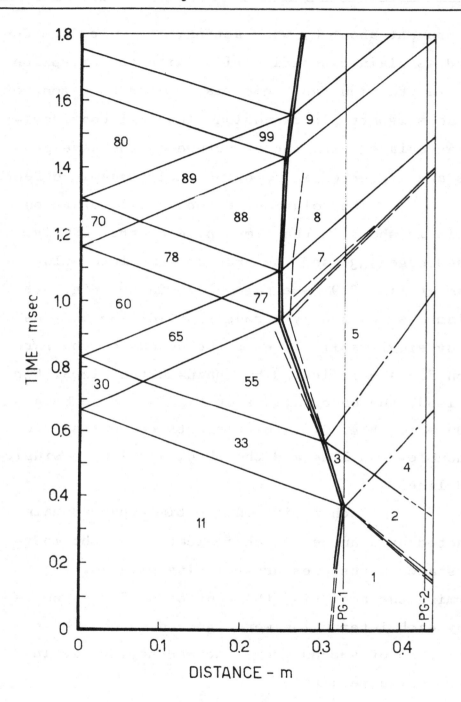

Fig. 3.8. Solution of the Flame-Shock Interaction Processes Recorded on Fig. 3.7. in the Space-Time Domain.

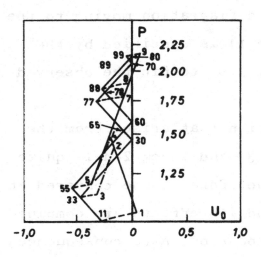

Fig. 3.9. Solution of the Flame-Shock
Interaction Processes Recorded on
Fig. 3.7 in the P-U Plane.

initiation process it has been established (12) that
the pressure of state 1 is 2.01 psi, and the corresponding particle velocity is 13 m/sec.

State 1 and the waves 1-2, 2-4, and
1-11 provide the initial conditions for the analysis.
For convenience, the reference conditions are taken as
those of the undisturbed mixture, so that state 1 is
slightly displaced from the origin in the $P - U$ diagram of Fig. 3.9.

Curve 1-2 in Fig. 3.9 represents the
locus of states attained behind the shock wave propagating to the left into medium 1, with the terminal
point, state 2, determined from the observed pressure

and wave velocity. Curve 1-11 represents the locus of
states created behind the deflagration moving to the
right into medium 1. State 11 is specified by the
deflagration polar of Fig. 3.6. to fit the observed
flame velocity.

The rarefaction that arises from the
collision between flame 3-33 and shock 2-4 is quite
weak. It is considered, therefore, to be centered at
the point of collision, and its effects are combined
with those of the first expansion. As a consequence,
there appears an extended regime of uniform state 5,
in agreement with experimental observations. Although
flame 3-33 propagates into the non-uniform regime
within the reflected rarefaction wave, the medium
ahead of it is assumed to have a constant state 3,
whose properties correspond to those behind the rarefaction produced by the first-shock collision. This
should be considered a fair approximation, especially
in view of the experimental evidence. The rarefaction
waves attributed to changes in flame shape propagate
in both directions, so that shock 11-33 is actually
followed by an expansion that tends to "smear out"
the subsequent interactions. For the purpose of analysis, however, state 33 is considered uniform, with
properties corresponding to those behind the trans-

mitted shock wave after it was overtaken by the
weaker rarefaction.

States 3 and 33 are determined on the
$P - U$ plane of Fig. 3.9 by closing the pentagon
whose remaining vertices are state points 11, 1 and 2.
A unique solution is obtained by requiring that the
velocity of the transmitted deflagration 3-33 match
the flame trace (world-line) on the schlieren record,
while point 33 is that behind the transmitted shock
11-33. The results of such a solution are quite approx
imate, since the pressure and velocity changes across
the deflagration are sensitive to small deviation in
flame velocity. However, an upper bound to state 3,
and hence state 33, is provided by state 5(which,to-
gether with state 55, is found in a similar fashion
as states 3 and 33), whose pressure was recorded by
both pressure transducers.

The results of the analysis are sum-
marized in Table 1 (see page 136). On the basis of
the time-space diagram of Fig. 3.8, pressure varia-
tions at the locations of the two pressure transducers are determined. The resulting profiles are
shown in Fig. 3.10 (see page 137) in comparison to
the experimental pressure records.

The satisfactory agreement between the

TABLE 3.1.

Solution of Flame-Shock Interactions of Fig. 3.7 Represented on Figs. 3.8, 3.9 and 3.10.

State	P	Δp (lb/in²)	A	u	Wave	M_i	$M_o = M_i A_i$	$u + M_o$	ω (m/sec)
1	1,04	0,075	1,0056	+0,028	0-1	—	0	0,028	13
2	1,393	0,76	1,05	-0,17	1-2	-1,14	-1,145	-1,117	-502
11	1,003	0,005	2,085	-0,29	1-11	—	0,078	0,106	48
3	1,133	0,26	1,02	-0,33	2-3	1	1,05	0,88	392
						1	1,02	0,69	310
33	1,119	0,23	2,19	-0,485	11-33	-1,05	-2,19	-2,48	-1120
					3-33	—	0,035	-0,295	-133
4	1,47	0,91	1,06	0,21	2-4	-1,024	-1,075	-1,25	-563
5	1,185	0,36	1,03	-0,375	4-5	1	1,06	0,85	
						1	1,03	0,655	+294
55	1,168	0,325	2,19	-0 56	5-55	—	0,040	-0,335	-150
					33-55	-1,02	-2,235	-2,720	-1225
30	1,447	0,865	2,24	0	33-30	1,13	2,53	2,045	920
65	1,514	0,995	2,25	0,08	30-65	-1,02	-2,29	-2,29	-1030
					55-65	1,135	2,55	1,99	896
60	1,576	1,115	2,25	0	65-60	1,02	2,29	2,21	995
7	1,77	1,495	1,09	-0,07	5-7	1,190	1,30	0,925	915
77	1,725	1,405	2,27	-0,33	7-77	—	0,07	0	0
					65-77	-1,065	-2,42	2,50	-1125
8	1,88	1,70	1,10	-0,025	7-8	1,03	1,12	1,05	472
88	1,84	1,62	2,28	-0,28	8-88	—	0,07	0,0450	22
					78-88	-1,01	-2,30	-2,555	1150
78	1,803	1,55	2,28	-0,255	77-78	1,02	2,32	1,99	896
					60-70	-1,06	-2,46	-2,46	1100
70	2,065	2,06	2,30	0	78-70	1,06	2,42	2,165	972
89	2,095	2,12	2,30	-0,03	88-89	1,06	2,42	2,14	962
					70-89	1,008	-2,32	-2,32	-1045
80	2,125	2,17	2,31	0	89-80	1,008	2,32	2,29	1030
9	2,15	2,22	1,12	0,08	8-9	1,06	1,165	1,14	512
99	2,125	2,17	2,32	0,06	9-99	—	0,03	0,11	50
					89-99	-1,007	-2,32	-2,35	-1060

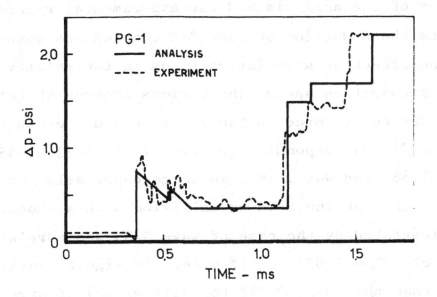

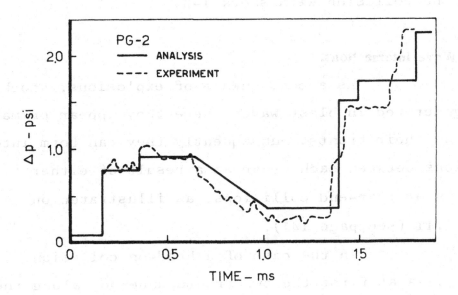

Fig. 3.10. Comparison between the Pressure Profiles Deduced from the Solutions of Figs. 3.8 and 3.9 and the Experimental Records of Fig. 3.7.

results of the analysis and the experimental records
permits the deduction of specific conclusions concern
ing the effect of wave interactions on the relative
flame propagation speed. The various changes of this
speed are represented in Table 1 by the different val
ues for M_o corresponding to waves 1-11, 3-33, 5-55,
7-77, 8-88, and 9-99. They indicate especially the
significance of the head-on collision with a shock,
as illustrated by the case of wave 3-33 whose relative
speed of $M_o = 0.035$ (or 15 m/sec) is significantly
lower than the $M_o = 0.078$ (or 35.1 m/sec) of wave
1-11 that represented the same flame front just before the collision with shock 1-2.

3. 11. Wave Intersections.

As a consequence of explosions, shocks
are generated in blast waves where they appear primarily as their fronts. Subsequently they can form inter
sections between each other as a result of either
head-on or rear-end collisions, as illustrated on
Fig. 3.11 (see page 141).

In the case of a head-on collision,
Fig. 3.11a at first the two fronts undergo, along the
line of centers, a normal interaction producing a
reflected shock. At the next instant quadruple shock

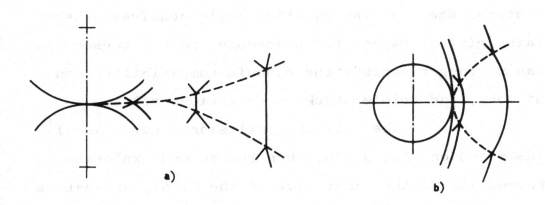

Fig. 3.11. Illustration of the formation of Shock Intersections in an Explosive Medium.

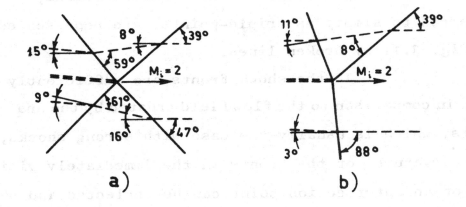

Fig. 3.12. Typical Shock Intersection Patterns.

(a) Quadruple intersection (equivalent to an asymmetric regular reflection).

(b) Inverse Mach intersection.

intersections are formed on both sides of the line of
centers. When the intersection angle acquires a certain critical value, the quadruple shock intersection
can no longer satisfy the dynamic compatibility conditions, and triple shock intersections set in.

The rear-end collision occurs, as illustrated on Fig. 3.11b, when the second explosion is
formed inside the blast wave of the first, or what is
of course equivalent, when an explosion occurs behind
a plane shock. In this case, at first the two shock
fronts merge along the line of centers, producing a
transmitted shock and a reflected rarefaction fan. At
the next instant, triple shock intersections are generated. The trajectories of these intersections,
referred to simply as triple-points, are represented
on Fig. 3.11 by broken lines.

If the shock fronts are sufficiently
thin in comparison to the flow field treated by the ana
lysis, which is usually the case with strong shocks,
the curvatures of the fronts in the immediately vicinity of an intersection point can be neglected and the
surrounding flow field considered to be essentially
the same as in the case of plane fronts. The various
wave configurations that are, under such circumstances, dynamically compatible are depicted on Fig.3.12.

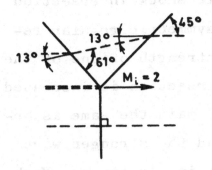

(c) Normal Mach intersection.

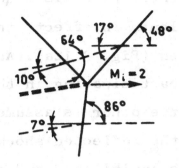

(d) Conventional Mach intersection

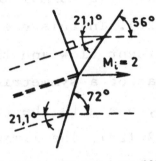

(e) Limiting case of the conventional Mach
intersection where reflected shock is normal
to local flow.

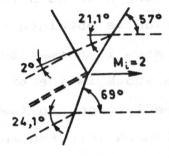

(f) Conventional "arrowhead" intersection.

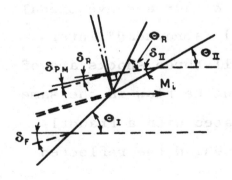

(g) Choked "arrowhead" intersection (sonic
flow behind reflected shock).

Fig. 3.12

Shown there first is a quadruple shock intersection
equivalent, in effect, to an asymmetric regular reflection (Fig. 3.12a). As the strength of one of the
waves participating in the intersection is increased,
while the other is assumed to remain the same as before, the reflected shock behind the stronger wave
becomes annihilated and one obtains an inverse Mach
intersection, shown next (Fig. 3.12b). Then, for still
higher strength of both incident waves, a normal Mach
intersection is obtained (Fig. 3.12c). In this case
the Mach stem is normal to the incident flow and the
reflected shock is identical to that of a symmetric
regular reflection. Following this, one gets the conventional Mach intersection (Fig. 3.12d). The limiting
case for this configuration is attained when the reflected shock becomes normal to its incident flow (Fig.
3.12e). After that one can have either a conventional
(Fig. 3.12f) or a choked (3.12g) "arrowhead" intersection, that is one in which all three shocks are of
the same family (i.e., they turn the flow in the same
sense), the latter being associated with a Prandtl-
Meyer expansion while the flow behind the reflected
shock´is locally sonic.

        The dynamic compatibility conditions
which a wave intersection has to satisfy stem from

the requirement that each component has to obey the
jump conditions for an oblique gasdynamic discontinuity, while the flow field generated by the intersection has to have uniform pressure and its particle
velocities with respect to the intersection point
must be mutually parallel. It is the latter condition
that gives rise to a slip-line discontinuity.

As a consequence of this requirement,
the key to the solution of most problems in this
domain is the relationship between the pressure and
the flow deflection angle which is called again, as
in the previous case, the wave polar. Plots of such
polars and their auxiliary diagrams for the evaluation
of the incidence angle, ϑ , and the Mach number, M_j
based on Eqs. (3.69), (3.70) and (3.71) for a shock
front in the case of a perfect gas with γ = 1.4 (i.
e., β = β_P = β_v = 1/6) are given on Figs. 3.13,
3.14 and 3.15 (see page 144 and 145). The correspond
ing polar and auxiliary plots for the Prandtl-Meyer
expansion are presented on Fig. 3.16 (see page 145).
Polars and auxiliary diagrams for a detonation and a
deflagration wave can be found in Ref. 12. With the
use of Figs. 3.13, 3.14 and 3.15, the various intersections depicted on Fig. 3.12 (except for case (g))
were solved graphically, as shown on Fig. 3.17.

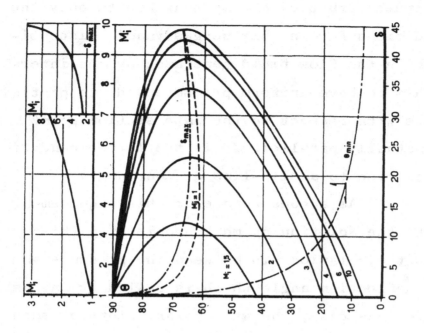

Fig. 3.14. Auxiliary $\theta = \theta(\delta)$ Plot for Fig. 3.13.

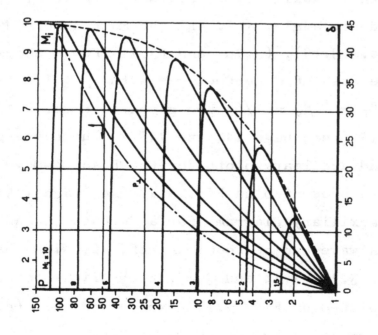

Fig. 3.13. Oblique Shock Polars for a Perfect Gas with $\gamma = 1.4$.

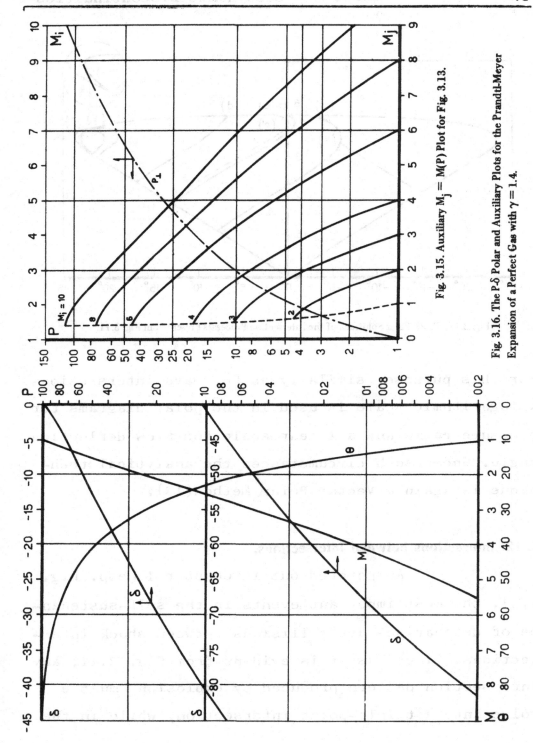

Fig. 3.15. Auxiliary $M_j = M(P)$ Plot for Fig. 3.13.

Fig. 3.16. The P-δ Polar and Auxiliary Plots for the Prandtl-Meyer Expansion of a Perfect Gas with $\gamma = 1.4$.

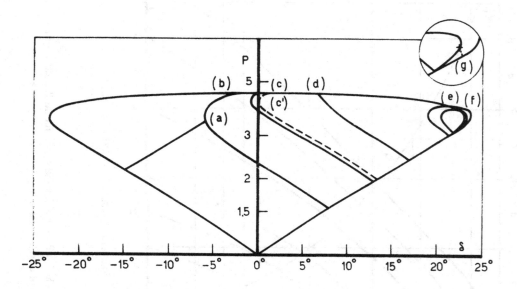

Fig. 3.17. Shock Polar Solutions of the Intersection Processes Presented in Fig. 3.11.

For this purpose, similarly as for wave intersections, a logarithmic scale is used in the polar diagrams for pressure ratio and a linear scale for flow deflection angle. Under such circumstances the analytical technique is again a Vector Polar Method(13).

3. 12. Interactions between Intersections.

As pointed out in Chapter 1 (esp. Fig. 1.6), the most important events in the self-sustenance of detonations are collisions between shock intersections. Since, as it is evident from Fig. 3.11, any intersection pattern produced by explosion must evolve into a triple-point intersection, while in the

most common case of a shock-induced explosion represented by Fig. 3.11b, such intersections are formed
right at the outset, only this case need be considered. Presented here are thus typical cases of, first,
a symmetric and, then, an asymmetric collision between
triple points(14).

A symmetric collision between triple-
point intersections is described schematically in Fig.
3.18 (see page 148) showing the wave configurations
immediately before and after this event. As a rule,
the intersection point "consumes" the weaker incident
shock, I. In order for a collision to occur, this
shock must be thus "consumed" from two sides, and the
collision takes place then at the moment when it becomes annihilated. Immediately after this, the two
stronger shocks, II, of the previous configuration
acquire the role of the weaker shocks, I', while either a new stronger shock, II', is formed "pushing"
the triple points away from the axis of symmetry, or
a regular reflection is obtained, as that of Fig.
3.12c, so that the resultant triple point "slides"
along the axis of symmetry. In the latter case a
Mach intersection will eventually be formed, as illustrated by Fig. 3.11a. Of particular significance to
the self-sustenance of detonations is here the fact

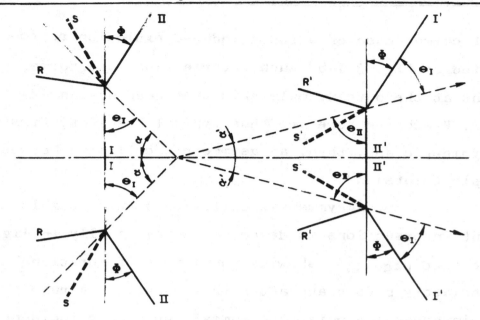

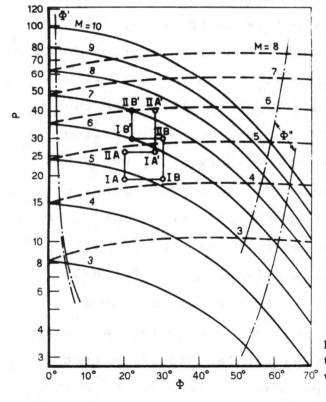

Fig. 3.18. Schematic Diagram of a Symmetric

Collision between Triple-Points.

Fig. 3.19. Solutions of Collision Processes Between Triple-Points on a Plane of Collision Invariants for a Perfect Gas with $\gamma = 1.4$.

that, as a result of this process, in both cases a
high pressure and temperature region associated with
a concentrated vortex is created in the immediate
vicinity of the collision point.

The wave system resulting from a sym-
metric collision is fully determined by the fact that
the weaker incident shock after collision was the
stronger shock before, while, due to the symmetry, the
newly formed stronger shock is perpendicular to the
axis of symmetry. The process of collision has, therefore, two invariants : the pressure ratio, P , across
the wave which remains unaffected by this process and
the angle between the two front shocks, Φ. Exploiting
the plane of collision invariants, one can thus cons-
truct a chart for the evaluation of the resultant
wave system. Figure 3.19 (see page 148) represents
such a chart for the case of a perfect gas with $\gamma =$
$= 1.4$. The wave systems that can be obtained after
collision are represented there by continuous lines
determined by taking for P the pressure ratio across
the weaker incident shock. The corresponding wave systems before collision are represented by broken lines;
these were obtained simply by taking for P the pressure ratio across the stronger shock of the same
intersection. The chain-dotted lines denoted by Φ'.

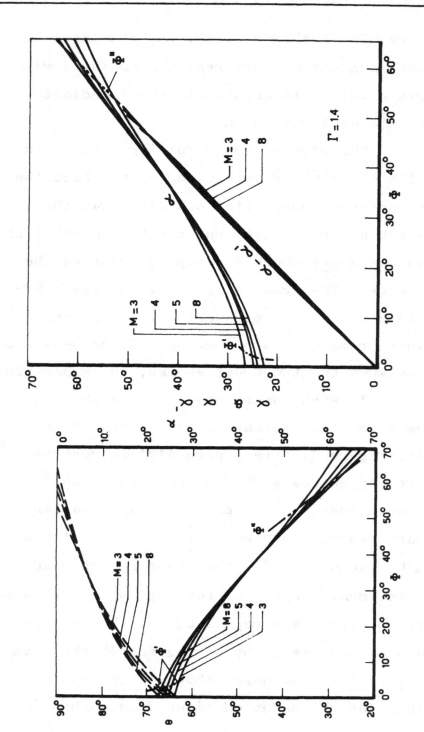

Fig. 3.21. Solutions of Collision Processes Corresponding to Fig. 3.19. Expressed in Terms of the Triple-Point Trajectory Angles.

Fig. 3.20. Solutions of Collision Processes Corresponding to Fig. 3.19. Expressed in Terms of Shock Incidence Angles.

represent the loci of the normal reflected shock solutions (case (e)) of Figs. 3.12 and 3.17), while
those denoted by Φ'' are the regular reflection limits
(case (c) of Figs. 3.12 and 3.17) beyond which the
angle of reflection, $\alpha' = 0$.

On the $P - \Phi$ plane of Fig. 3.19, each
point of intersection between a continuous line and
a broken line represents a solution of a collision
process, specifying fully the wave systems immediately before and after this event. Once the solution in
terms of P and M_i is known, the angles between the
trajectories of the triple points can be determined
from the auxiliary $\vartheta = \vartheta(\delta)$ plot , Fig. 3.15, as,
by reference to Fig. 3.18,

$$(3.101) \quad \alpha = \frac{\pi}{2} - \vartheta_I \quad \text{while} \quad \alpha' = \frac{\pi}{2} - \vartheta'_{II} = \frac{\pi}{2} - \Phi - \vartheta'_I$$

Thus, making use of Fig. 3.19 and the
plots of Figs. 3.14 and 3.15, the angles between the
trajectories of the triple point intersections before
and after symmetric collisions were evaluated for a
set of incident flow Mach numbers, and the results
are given on Figs. 3.20 and 3.21 (see page 157, 158).
From the latter one should note that $\alpha - \alpha'$ becomes
negligibly small as Φ tends to zero which corres-

ponds to $M_i = 1$, the condition which is attained when the reflected shock becomes reduced to a characteristic. Moreover, as it appears there, the relationship between θ_I and Φ is practically independent of the incident flow Mach number, especially if its value is high enough. Since this means that in Eqs. (3.101) $\theta_I = \theta_I'$ one has therefore, in this case, quite a simple rule for estimating, with good approximation, the change in the angle between the trajectories of the triple-points due to their collision, namely,

$$\alpha - \alpha' \cong \Phi \qquad (3.102)$$

An example of wave configurations associated with an asymmetric collision is represented by Fig. 3.22 (see page 153). The wave system before collision shares the incident shock, so that $P_{IA} = P_{IB}$ The resultant wave system is determined by four conditions which are imposed by the requirement that the two stronger shocks of the initial wave system are unaffected by the collision, acquiring to roles of the weaker incident shocks after this event, while the newly formed stronger shock is common to both the resultant intersections A' and B'. One has, therefore, the following equalities : (see page 154).

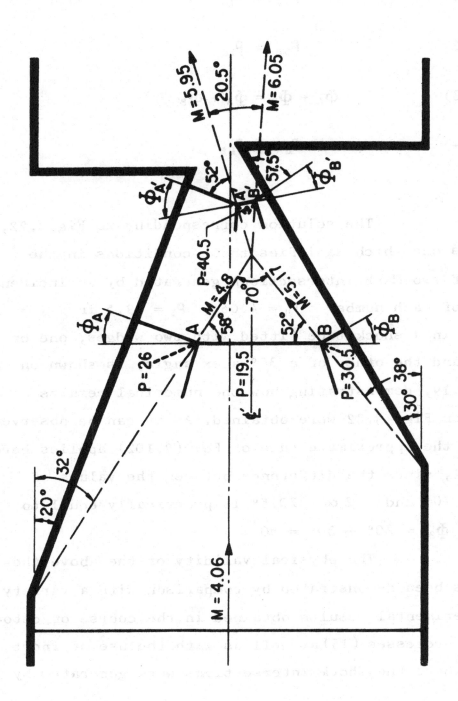

Fig. 3.22. An example of an Asymmetric Collision between Triple-Points.

(1) $$P_{IIA} = P_{IA'}$$

(2) $$P_{IIB} = P_{IB'}$$

(3) $$\Phi_A + \Phi_B = \Phi_{A'} + \Phi_{B'}$$

(4) $$P_{IIA'} = P_{IIB'} \quad .$$

The solution corresponding to Fig.3.22, that is one which satisfies these conditions in the case of two Mach intersections generated by an incident shock of Mach number $M_i = 4.06$ ($P_I = 19.5$ if $\gamma = 1.4$) in a shock tube fitted with two wedges, one of a 20° and the other of a 30° apex angle, is shown on Fig. 3.19, demonstrating how the numerical results given on Fig. 3.22 were obtained. As it can be observed there, the approximate rule of Eq. (3.102) applies here as well, since the difference between the values of $2\alpha = 70°$ and $2\alpha' = 20.5°$ is practically equal to $\Phi_A + \Phi_B = 20° + 30° = 50°$.

The physical validity of the above theory has been demonstrated by comparison with a variety of experimental results obtained in the course of detonation processes (15) as well as with the use of inert gases where the shock intersections were generated by

means of wedges in shock tubes as in the case of
Fig. 3.22. Selected for illustration here were, in
fact, the records corresponding to this case. They
were obtained in a shock tube 1-3/4" x 1-3/4" in
cross-section filled initially with nitrogen at room
temperature and a pressure of 15 mm Hg, the conditions
for which the assumption that the substance behaves
as a perfect gas with $\gamma = 1.4$ is indeed very good.
Stroboscopic laser-schlieren records of the wave systems obtained when the incident shock Mach number was
4.06 are given by Fig. 3.23 (see page 157) including
a record of the corresponding smoke film traces of
the triple point traiectories. Their agreement with
Fig. 3.22 is indeed most satisfactory.

In ref. [14] a similar analysis was
presented for triple-point traces recorded in the
case of a condensed explosive. The medium was a nitro-
methane-acetone mixture for which the Rankine-Hugoniot
curve on the pressure-specific volume plane is not
represented by a rectangular hyperbola. In spite of
this, the method of analysis was essentially the same
as that presented here for the case of a perfect gas.

Again, a satisfactory agreement between
the analytical results of triple-point collisions and the
experimental traces was obtained. Unlike the gaseous

medium, however, it turned out that the most likely
wave pattern to occur in the condensed explosive was
an "arrowhead" intersection rather than a Mach reflection. This was associated with the large change in Γ
across the discontinuity. As a consequence, case (g)
of Figs. 3.12 and 3.16, rather than case (f), was obtained then for intersections at small angles Φ, and
the limit of "arrowhead" intersections on the plane
of collision invariants, Φ', was of an order of 15°,
rather than 3° - 5° as it appears on Fig. 3.18 for a
perfect gas.

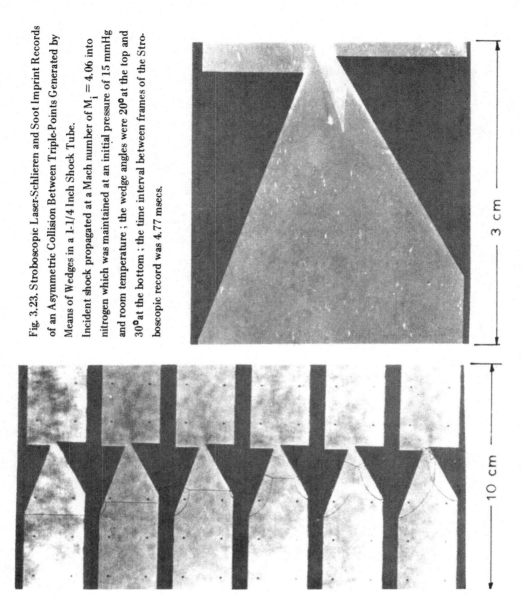

Fig. 3.23. Stroboscopic Laser-Schlieren and Soot Imprint Records of an Asymmetric Collision Between Triple-Points Generated by Means of Wedges in a 1-1/4 Inch Shock Tube.

Incident shock propagated at a Mach number of $M_i = 4.06$ into nitrogen which was maintained at an initial pressure of 15 mmHg and room temperature ; the wedge angles were $20°$ at the top and $30°$ at the bottom ; the time interval between frames of the Stroboscopic record was 4.77 msecs.

References.

1. Courant, R. and Friedrichs, K.O., <u>Supersonic Flow and Shock Waves</u>, Interscience Publishers, Inc., New York, XVI + 464 pp., 1948.

2. Oswatitsch, Klaus, <u>Gas Dynamics</u>, Academic Press Inc., New York, XV + 610 pp. and 3 charts, 1956.

3. Rudinger, George, <u>Wave Diagrams for Nonsteady flow in Duts</u>, D. Van Nostrand Company, Inc., New York, XI + 278 pp., 1955.

4. Glass, I.I. and Hall, J. Gordon, "Handbook of Supersonic Aerodynamics-Section 18-Shock Tubes, "Bureau of Ordnance, Department of the Navy, Navord Report 1488, <u>6</u>, XXXVIII + 604 pp., 1959.

5. Shchelkin, K.I. and Troshin, Ya. K., <u>Gazodynamika Gorenia</u> (Gasdynamics of Combustion), Izdatelstvo Akademii Nauk SSSR, Moscow, 255 pp., 1963 ; Transl. Mono Book Corp., Baltimore, VI + 222 pp., 1965.

6. Zel'dovich, Ya. B. and Rayzer, YU. P., Fizika Udarnykh Voln i Vysokotemperaturnykh Hidrodinamicheskikh Yavlenyi (Physics of Shock Waves and High-Temperature Hydrodynamic Phenomena), Gos. Izd. Fiz. Mat. Literatury, 686 pp., Moscow, 1963 ; (Transl., Edited by W. D. Hayes and R.F. Probstein), Academic Press, New York, I and II, XXIII + XXIV + 916 pp., 1966-67.

7. Jouguet, E., "Mécanique des Explosifs," O. Doin et Fils, Paris, XX + 516 (esp. § 193,

pp. 278 - 279), 1917.

8. Becker, R., "Stosswelle und Detonation," Zeitschrift für Physik, $\underline{8}$, 321 - 362 (esp.
p. 352), 1922.

9. Langweiler, H., "Beitrag zur Hydrodynamischen Deto-
nationstheorie". Zeitschrift für tech-
nische Physik, $\underline{19}$, 9, 271 - 283, 1938.

10. Murnaghan, F. D., Finite Deformation of an Elastic
Solid, Wiley, New York, p. 68, 1951.

11. Oppenheim, A. K., Urtiew, P. A. and Laderman, A.J.,
"Vector Polar Method for the Evaluation
of Wave Interaction Processes",
Archiwum Budowy Maszyn, $\underline{XI}$, 3, 441-495,
1964.

12. Laderman, A.J., Urtiew, P.A. and Oppenheim, A.K.,
"Gasdynamic Effects of Shock-Flame
Interactions in an Explosive Gas",
AIAA Journal, $\underline{3}$, 5, pp. 876-883, May
1965.

13. Oppenheim, A.K., Smolen, J.J. and Zajac, L.J.,
"Vector Polar Method for the Analysis
of Wave Intersections," Combustion and
Flame, $\underline{12}$, 1, 63-76, 1968.

14. Oppenheim, A.K., Smolen, J.J., Kwak, D., Urtiew,
P.A., "On the Dynamics of Shock Intersections," Fifth Symposium(International) on Detonation, Pasadena, California,
17 pp., 1970.

15. Strehlow, R.A., "Gas Phase Detonations : Recent
Developments" Combustion and Flame, $\underline{12}$,
2, pp. 81 - 101, April 1968.

Chapter 4.
Blast Waves.

The Flow Fields of Explosions.

4. 1. Introduction.

Blast waves are, essentially, non-steady flow fields generated by explosions. The development of the blast wave theory has been spearheaded by the interest in the effects of atom bombs, having been founded in the nineteen-forties upon such notable contributions as those of Taylor(1,2)[*], von Neumann(3,4), Sedov(5,6) and Stanyukovich(7,8).

In the intervening years, the blast wave theory received a good deal of attention, as exemplified by a substantial number of publications. Among the comprehensive expositions one has, for instance, the book of Korobeinikov, Mil'nikova and Ryazanov (9) and the monograph of Sakurai(10), while authoritative chapters on this subject can be found in the text of Courant and Friedrichs (11) and in the

---

*) Numbers in parentheses denote references listed at the end of the chapter.

book of Zel'dovich and Raizer(12). Of the original
contributions, one may note, for instance, the papers
of Brode(13,14) on numerical analysis;of Grigorian(15)
on self-similar piston-driven waves ; of Oshima (16)
on quasi-similar solutions ; of Korobeinikov and
Chuskin (17) on numerical solutions for blast waves
expanding into an atmosphere of finite counter-pressure ; of Levin and Chernyi(18) on asymptotic solutions for waves bounded by detonation fronts ; of Bishimov, et al(19) and Korobeinikov (20) on blast waves
in explosive gases ; of Laumach and Probstein(21) on
the point explosion in an exponential atmosphere and
of Lee(22) on implosions in a detonating gas ; to cite
just a few in order to portray the scope of interest
in this field of study.

In the prevailing number of cases, the
problems were formulated in the Eulerian frame of
reference and, being concerned mainly with the effects
of explosions,they dealt primarily with constant energy,
decaying blast waves. Moreover, the applicability of
fundamental equations used for this purpose had been
restricted by the assumption, introduced usually right
at the outset, that the substance behaves as a perfect
gas with constant specific heats. In this type of
problems, the main purpose of the analysis was the

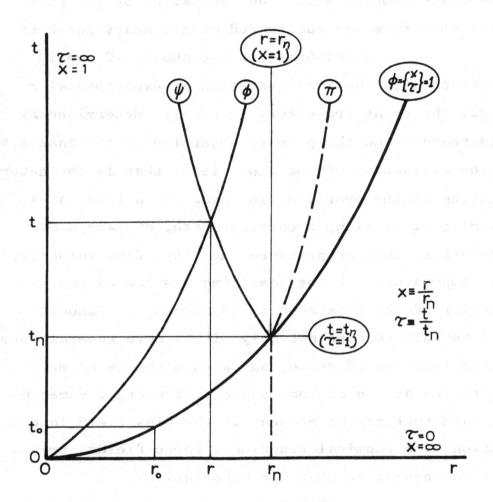

Fig. 4.1. Blast wave coordinates.

determination of the trajectory of the front in the
time-space domain, while the evaluation of the flow
field structure was considered of secondary interest.

In contrast to the above, if one is
concerned with the interpretation of experimental records, the front trajectory is usually determined by
measurement, and the primary objective of the analysis
is the evaluation of the flow field, that is the deter
mination of the profiles, in space or in time, at a
fixed point or along a particle path, of gasdynamic
parameters, such as pressure, density, flow velocity,
and temperature. At the same time, as far as the
behavior of the substance is concerned, one should
then be able to cope not only with a more general form
of the equation of state, but also with the effects
due to the action of some sources of energy, momentum
and mass that may be present in the flow field in connection with chemical reactions, force fields, transport processes, or phase transformations.

For this purpose, therefore, the fundamental conservation equations are formulated here with
out restrictions imposed by a specific form of an equa
tion of state, while proper account is taken, by means
of source terms, of any other effects, that, besides
the inertial terms occupying the dominant position in

these equations, can affect the flow. Taking advantage
of the fact that the blast wave can be considered, in
general, as a spatially one-dimensional flow field
whose non-steady behavior can be, consequently, regarded as a function of just two independent variables,
two generalized blast wave coordinates are introduced.
This renders the equations a comprehensive character, in
that they can be then used directly to determine either space or time profiles of the dependent gasdynamic parameters in the Eulerian system, or time profiles in the Lagrangian frame of reference. In this
respect we have been influenced by a similarly general
manner in which Jeffrey and Taniuiti(23) treated the
wave propagation theory, especially in connection with
their introduction of the generalized curvilinear wave
coordinates for subsequent use as the characteristics
in the Cauchy problem.

        The fundamental equations are thus, ac-
cordingly, transformed, and then, they are reduced to
a more concise, autonomous, form for an arbitrary
equation of state relating the internal energy with
pressure and density. The formulation of the problem
is, finally, completed by the specification of the
boundary conditions imposed by a gasdynamic discontinuity, which, consistently with the rest of our

treatment, are expressed in a manner that is virtually
unrestricted by a particular from of the equation of
state.

To facilitate the comparison between the
present exposition and other publications in this
field, a list of the salient non-dimensional parameters used by various authors of blast wave theory is
given in Table 4.1. (See page 167).

4. 2. Conservations Principles.

The non-steady, one-dimensional flow
field is a function of just two independent variables,
the time, t , and the space coordinate, r . If the
stream cross-section is time-independent, as in the
case of a blast wave, the conservation equations
governing the flow can be expressed most conveniently
in the following divergence form :

$$\frac{\partial}{\partial t} a_k + \frac{\partial}{\partial r} (u\, b_k) = \dot{c}_k \qquad (4.1)$$

where a_k, b_k and $\dot{c}_k$ can be regarded as vector quantities so that :

$$k = \begin{cases} 0 \text{ for the conservation of mass} \\ 1 \text{ for the conservation of momentum} \\ 2 \text{ for the conservation of energy.} \end{cases}$$

TABLE 4.1.

COMPARISON BETWEEN BASIC NON-DIMENSIONAL PARAMETERS USED BY VARIOUS AUTHORS OF BLAST WAVE THEORY.

Quantities / References	Relative Space Coordinate (x)	Decay Coefficient (λ)	Velocity Modulus (μ)	Relative Particle Velocity (f)	Density Ratio (h)	Pressure Ratio (g)	Reduced Particle Velocity (F)	Reduced Velocity of Sound (Z)
Present Exposition	$\dfrac{r}{r_1}$	$\dfrac{d \ln g}{d \ln r_1}$	$\dfrac{d \ln \xi}{d \ln \eta}$	$\dfrac{u}{w_1}$	$\dfrac{\varrho}{\varrho_0}$	$\dfrac{P}{\varrho_0 w_1^2}$	$\dfrac{\tau}{x} f$	$\left(\dfrac{\tau}{x}\right)^2 \dfrac{g}{h}$
Taylor 1950	η	—	—	Φ	ψ	$\left(\dfrac{1}{\gamma}\right) f$	—	—
Courant & Friedrichs 1948	$\dfrac{\xi}{\xi_1}$	—	α	—	$(x^{\alpha_0}) \dfrac{\Omega}{\Omega_0}$	$(x^{\alpha_0}) \dfrac{P}{\Omega_0}$	U	$\dfrac{1}{\gamma} c^2$
Sedov 1957	λ	—	δ	$(f_1) f$	$(h_1) g$	$(g_1) h$	$\dfrac{1}{\mu}$	$\dfrac{1}{\gamma \mu^2} F$
Korobeinikov et al. 1961	λ	$\dfrac{d \ln g}{d \ln r_2}$	δ_1	$(f_1) f$	$(h_1) g$	$(g_1) h$	$(f_1) F$	—
Sakurai 1965	x	λ	—	f	h	$\left(\dfrac{1}{\gamma}\right) g$	F	—
Zeldovic & Raizer 1966	ξ	—	α	v	g	π	$\left(\dfrac{1}{\mu}\right) v$	$\dfrac{1}{\gamma \mu^2} z$

Note : Symbols in parentheses are those of the present exposition.
All other references correspond to $\tau = 1$.

Explicitly, the components of these vectors are specified as follows :

k	$\dfrac{a_k}{\rho r^j}$	$\dfrac{b_k}{\rho r^j}$	$\dfrac{\dot{c}_k}{\rho r^j}$
0	1	1	Ω_M
1	u	$u + \dfrac{p}{\rho u}$	$\dfrac{jp}{\rho r} + \Omega_F$
2	$e + \dfrac{u^2}{2}$	$e + \dfrac{u^2}{2} + \dfrac{p}{\rho}$	Ω_E

$$(4.2)$$

where :

$$j \equiv \frac{d\ln A}{d\ln r} = \begin{cases} 0 \text{ for plane symmetrical flow} \\ 1 \text{ for line \ \ symmetrical flow} \\ 2 \text{ for point symmetrical flow} \end{cases}$$

In the above A is the flow cross-sec-
tion area, ρ –density, p –pressure, u –flow velocity,
e – specific internal energy, Ω_M –rate of mass supplied from a source per unit mass of the flowing substance, Ω_F –body or dissipative force, and Ω_E –rate
of energy deposited or dissipated per unit mass of
substance.

4. 3. Blast Wave Transformation.

The various blast wave coordinates in
the time-space domain are depicted on Fig. 4.1 (see
page 163). The origin of the system is, as a rule, at

$t = 0$ and $r = 0$. The coordinates of the front are t_n and r_n and the most convenient non-dimensional coordinates of the flow field are defined with their respect as the ratios $\tau = \dfrac{t}{t_n}$ and $x = \dfrac{r}{r_n}$. The trajec tory of the front in the time-distance domain can, in turn, be expressed in terms of the non-dimensional variables $\eta = \dfrac{t_n}{t_0}$ and $\xi = \dfrac{r_n}{r_0}$ where t_0 and r_0 are the co-ordinates of an arbitrary reference point on the tra jectory. The non-dimensional parameter describing its slope, that acquired most popularity in the literature (see e.g. Sakurai (10)), is $y \equiv \dfrac{a_a^2}{w_n^2}$ (denoted by q in the text of Korobeinikov et al.(9)), where a_a is the local velocity of sound in the undisturbed medium in-to which the wave propagates and w_n is the front ve-locity. The particle path is presented on Fig. 4.1 as a line of $\Pi(r,t) = $ const, so that, the flow velocity is $u = (\partial r/\partial t)_\pi$. In non-dimensional form the latter is expres sed in terms of $f = \dfrac{u}{w_n}$. Since the blast wave represents, in essence, a two-dimensional flow field in the time-space domain, the fundamental equations can be render-ed a comprehensive character by the introduction of two generalized blast wave coordinates, $\psi(r,t)$ and $\varphi(r,t)$.   The first is defined in such manner that is describes the motion of the front and, con-sequently, it can be referred to as the "front coordinate". As shown on Fig. 4.1, its main feature is the fact that a ψ = const. line passes through a point on the front trajectory. In assigning to it a particular value, it can be considered as equal to ξ or η or y.

The second is meant to describe the flow field and it can be referred to, therefore, as the "field coordinate". The salient feature of this coordinate is the fact that, in contrast to the former, the ϕ = const. lines cannot cross the front trajectory, except at the origin, while the front trajectory itself belongs, as a rule, to this family. In assigning to it a particular value, it can be regarded as equal to x or τ or f . In the first two cases the front trajectory is represented by a ϕ = 1 line.

Now, as demonstrated on Fig. 4.1, in order to obtain space profiles in the Eulerian frame of reference, the front coordinate is prescribed so that the $\psi(r,t)$ = const lines coincide with lines of $t = t_n$ for which τ = 1, while the field coordinate $\phi = x$. The time profiles in the same frame of reference correspond to the case where the $\psi(r,t)$ = const lines coincide with lines $r = r_n$ for which x = 1, while $\phi = \tau$. Finally, for the time profiles in the Lagrangian frame of reference, the ψ = const lines coincide with the particle paths, that is curves of $\Pi(r,t) = \text{const}$, so that $\left(\dfrac{\partial r}{\partial t}\right)_\psi = u$ while $\phi = \tau$. The fundamental equations are then transformed by noting that, for any function $F(r,t)$,

$$\frac{\partial F}{\partial t} = \frac{J(r,F;\psi,\phi)}{J(r,t;\psi,\phi)} \quad \text{and} \quad \frac{\partial F}{\partial r} = \frac{J(F,t;\psi,\phi)}{J(r,t;\psi,\phi)}$$

where J denotes the Jacobian, or

$$J(u,v;x,y) \equiv \frac{\partial(u,v)}{\partial(x,y)} = \frac{\partial u}{\partial x}\frac{\partial v}{\partial y} - \frac{\partial v}{\partial x}\frac{\partial u}{\partial y} .$$

For a continuous flow field, the functions $\psi(r,t)$ and $\phi(r,t)$ are continuous to, at least, the first derivative, so that

$$\frac{\partial^2 r}{\partial\psi\,\partial\phi} = \frac{\partial^2 r}{\partial\phi\,\partial\psi} \quad \text{and} \quad \frac{\partial^2 t}{\partial\psi\,\partial\phi} = \frac{\partial^2 t}{\partial\phi\,\partial\psi} \quad (4.3)$$

As a consequence of this property, one obtains the following transformation formulae :

and

$$\left.\begin{array}{l} \dfrac{\partial F}{\partial t} = - \dfrac{\dfrac{\partial}{\partial\psi}\left(F\dfrac{\partial r}{\partial\phi}\right) - \dfrac{\partial}{\partial\phi}\left(F\dfrac{\partial r}{\partial\psi}\right)}{J(r,t;\psi,\phi)} \\[4ex] \dfrac{\partial F}{\partial r} = \dfrac{\dfrac{\partial}{\partial\psi}\left(F\dfrac{\partial t}{\partial\phi}\right) - \dfrac{\partial}{\partial\phi}\left(F\dfrac{\partial r}{\partial\psi}\right)}{J(r,t;\psi,\phi)} \end{array}\right\} \quad (4.4)$$

With the above, Eq. (4.1) becomes :

$$(4.5) \quad -\frac{\partial}{\partial \psi}\left(a_k\frac{\partial r}{\partial \phi} - ub_k\frac{\partial t}{\partial \phi}\right) + \frac{\partial}{\partial \phi}\left(a_k\frac{\partial r}{\partial \psi} - ub_k\frac{\partial t}{\partial \psi}\right) = \dot{c}_k \mathfrak{J}(r,t;\psi,\phi).$$

4. 4. Conservation Equation in Non-Dimensional Form.

Now, the fundamental equations can be non-dimensionalized by the introduction of the following variables :

$$(4.6) \quad A_k \equiv \frac{a_k}{\rho_a r_n^j w_n^k} \; ; \quad B_k \equiv \frac{b_k}{\rho_a r_n^j w_n^k} \; ; \quad C_k \equiv \frac{\dot{c}_k}{\rho_a r_n^{j-1} w_n^{k+1}} \; .$$

As indicated earlier, the field coordinates are given by

$$(4.7) \quad x(\psi,\phi) = \frac{r}{r_n} \; ; \quad \tau(\psi,\phi) = \frac{t}{t_n} \; ; \quad f(\psi,\phi) = \frac{u}{w_n}$$

while the front coordinates are represented by

$$(4.8) \quad \xi(\psi) = \frac{r_n}{r_o} \; ; \quad \eta(\psi) = \frac{t_n}{t_o} \; ; \quad y(\psi) \equiv \frac{a_a^2}{w_n^2} \; .$$

It is convenient to introduce, at this stage, the logarithmic derivatives of the front properties with respect to the space coordinate. These will be denoted here by primes, so that, in general, one has

$$\psi_n' = \frac{d \ln \psi}{d \ln r_n} = \frac{d \ln \psi}{d \ln \xi} \qquad (4.9)$$

while, with reference to density immediately ahead of the front,

$$\rho_a' = \frac{d \ln \rho_a}{d \ln r_n} = \frac{d \ln \rho_a}{d \ln \xi} . \qquad (4.10)$$

Taking $\psi = -w_n^2$ Eq. (4.9) yields the definition of the well-known "decay coefficient"

$$\lambda = -2 \frac{d \ln w_n}{d \ln \xi} = \frac{d \ln y}{d \ln \xi} - 2 \frac{d \ln a_a}{d \ln \xi} . \qquad (4.11)$$

Associated more directly with the front velocity is the "velocity modulus" defined here, conversely as :

$$\mu = \frac{d \ln \xi}{d \ln t_n} = \frac{d \ln r_n}{d \ln t_n} = \frac{w_n t_n}{r_n} . \qquad (4.12)$$

The above two parameters are interrelated by the equation

$$\mu' = \frac{d \ln \mu}{d \ln \xi} = \frac{1}{\mu} - \frac{\lambda + 2}{2} \qquad (4.13)$$

which follows immediately from their definitions.

Using the above non-dimensional para-
meters, the three terms of Eq. (4.5) can be expressed;
respectively, as follows :

$$(4.14) \begin{cases} a_k\dfrac{\partial r}{\partial \phi} - ub_k\dfrac{\partial t}{\partial \phi} = \rho_a r_n^{j+1}\dfrac{w_n^k}{\phi}\left[A_k x\dfrac{\partial \ln x}{\partial \ln\phi} - fB_k\tau\mu\dfrac{\partial \ln\tau}{\partial \ln\phi}\right] \\[2ex] a_k\dfrac{\partial r}{\partial \psi} - ub_k\dfrac{\partial t}{\partial \psi} = \rho_a r_n^{j+1}\dfrac{w_n^k}{\psi}\left[A_k x\dfrac{\partial \ln(x\xi)}{\partial \ln\psi} - fB_k\tau\mu\dfrac{\partial \ln(\tau\eta)}{\partial \ln\psi}\right] \\[2ex] \qquad = \rho_a r_n^{j+1}\dfrac{w_n^k}{\psi\psi_n'}\left[A_k x\left(\psi_n'\dfrac{\partial \ln x}{\partial \ln\psi}+1\right) - fB_k\tau\left(\mu\psi_n'\dfrac{\partial \ln\tau}{\partial \ln\psi}+1\right)\right] \\[2ex] \dot{c}_k J(r,t;\psi,\phi) = -\rho_a r_n^{j+1}\dfrac{w_n^k}{\phi\psi}C_k x\tau\left[\dfrac{\partial \ln x}{\partial \ln\phi}\mu\dfrac{\partial \ln(\tau\eta)}{\partial \ln\psi} - \mu\dfrac{\partial \ln\tau}{\partial \ln\phi}\dfrac{\partial \ln(x\xi)}{\partial \ln\psi}\right] \\[2ex] \qquad = -\rho_a r_n^{j+1}\dfrac{w_n^k}{\phi\psi\psi_n'}C_k x\tau\left[\dfrac{\partial \ln x}{\partial \ln\phi}\left(\mu\psi_n'\dfrac{\partial \ln\tau}{\partial \ln\psi}+1\right) - \mu\dfrac{\partial \ln\tau}{\partial \ln\phi}\left(\psi_n'\dfrac{\partial \ln x}{\partial \ln\psi}+1\right)\right]. \end{cases}$$

Finally, noting that

$$(4.15) \qquad \psi_n'\frac{\partial}{\partial \ln\psi} = \frac{\partial}{\partial \ln\xi} = \frac{1}{\mu}\frac{\partial}{\partial \ln\eta} = -\frac{\lambda}{2}\frac{\partial}{\partial \ln w_n}$$

depending on whether $\psi = \xi, \eta$ or w_n, and selecting, for
the sake of simplicity, ξ to represent, in accordance
with the above, the front coordinate, Eq. (4.5) can be
written as follows :

$$(4.16) \qquad \left(\rho_a'+j+1-\frac{\lambda k}{2}\right)M_k + \frac{\partial M_k}{\partial \ln\xi} - \frac{\partial N_k}{\partial \ln\phi} = K_k$$

where

$$M_k \equiv A_k x \frac{\partial \ln x}{\partial \ln \phi} - f B_k \tau \mu \frac{\partial \ln \tau}{\partial \ln \phi}$$

$$N_k \equiv A_k x \left(\frac{\partial \ln x}{\partial \ln \xi} + 1 \right) - f B_k \tau \left(\mu \frac{\partial \ln \tau}{\partial \ln \xi} + 1 \right)$$

$$K_k \equiv C_k x \tau \left[\frac{\partial \ln x}{\partial \ln \phi} \left(\mu \frac{\partial \ln \tau}{\partial \ln \xi} + 1 \right) - \mu \frac{\partial \ln \tau}{\partial \ln \phi} \left(\frac{\partial \ln x}{\partial \ln \xi} + 1 \right) \right].$$

Explicitly, the transformed vector components of the conservation quantities can be expressed in terms of the following non-dimensional parameters

$$h \equiv \frac{\rho}{\rho_a} \quad, \quad g \equiv \frac{p}{\rho_a w_n^2} \quad \text{and} \quad \sigma \equiv \frac{e}{w_n^2}$$

while the source terms are represented by

$$\omega_i = \Omega_i \frac{r_h}{w_n^{k+1}} \tag{4.17}$$

where, respectively, $i = M, F, E$ for $k = 0, 1, 2$. Thus one obtains

k	$\dfrac{A_k}{hx^j}$	$\dfrac{B_k}{hx^j}$	$\dfrac{C_k}{hx^j}$
0	1	1	ω_M
1	f	$f + \dfrac{g}{fh}$	$\dfrac{jg}{xh} + \omega_F$
2	$\sigma + \dfrac{f^2}{2}$	$\sigma + \dfrac{f^2}{2} + \dfrac{g}{h}$	ω_E

$$\tag{4.18}$$

4. 5. Equation of State.

Equations (4.16) represents a system of three equations of two independent variables : ζ and Φ, and four dependent variables : f, h, g and σ The set is completed by the equation of state that expresses the internal energy in terms of pressure and density. In order to keep our exposition as compre hensive as possible, we introduce at this stage only the postulate that such relation exists without specifying its particular form. This can be done without restriction to any substance in thermodynamic equilibrium or any whose thermodynamic behavior could be considered as having not more than two degrees of freedom.

Thus, since $e = e(p, \varrho)$ it follows that

$$(4.19) \qquad de = \left(\frac{\partial e}{\partial \ln p}\right)_{\varrho} d\ln p + \left(\frac{\partial e}{\partial \ln \varrho}\right)_{p} d\ln \varrho \ .$$

The two partial derivatives can be expressed in a more convenient form by noting that from the above :

$$(4.20) \qquad \left(\frac{\partial e}{\partial \ln p}\right)_{s} = \left(\frac{\partial e}{\partial \ln p}\right)_{\varrho} + \left(\frac{\partial e}{\partial \ln \varrho}\right)_{p} \left(\frac{\partial \ln \varrho}{\partial \ln p}\right)_{s}$$

while for a process at constant entropy one has in general :

$$\left(\frac{\partial e}{\partial \ln p}\right)_s = \frac{p}{\varrho} \left(\frac{\partial \ln \varrho}{\partial \ln p}\right)_s \qquad (4.21)$$

Hence it follows that

$$\left(\frac{\partial e}{\partial \ln p}\right)_\varrho = \left[\frac{p}{\varrho} - \left(\frac{\partial e}{\partial \ln \varrho}\right)_p\right]\left(\frac{\partial \ln \varrho}{\partial \ln p}\right)_s . \qquad (4.22)$$

In terms of the non-dimensional isentropic compressibility, or velocity of sound, modulus, introduced by Eq. (3.5)

$$\Gamma = \left(\frac{\partial \ln p}{\partial \ln \varrho}\right)_s$$

and a non-dimensional internal energy factor

$$\varkappa = -\frac{\varrho}{p}\left(\frac{\partial e}{\partial \ln \varrho}\right) \qquad (4.23)$$

Eq. (4.22) can be expressed as

$$\left(\frac{\partial e}{\partial \ln p}\right)_\varrho = \frac{\varkappa + 1}{\Gamma}\frac{p}{\varrho} . \qquad (4.24)$$

Thus Eq. (4.19) becomes :

$$(4.25) \qquad de = \frac{\varkappa+1}{\Gamma} \frac{p}{\rho} \, d\ln p - \varkappa \frac{p}{\rho} \, d\ln \rho$$

Taking partial derivatives of the above with respect to the field coordinate, ϕ , and the front coordinate, ξ , respectively, substituting at the same time the non-dimensional wave variables for the state parameters, one obtains finally :

$$(4.26a) \qquad \frac{\partial \sigma}{\partial \ln \phi} = \frac{\varkappa+1}{\Gamma} \frac{g}{h} \frac{\partial \ln g}{\partial \ln \phi} - \varkappa \frac{g}{h} \frac{\partial \ln h}{\partial \ln \phi}$$

and

$$(4.26b) \qquad -\lambda \sigma + \frac{\partial \sigma}{\partial \ln \xi} = \frac{\varkappa+1}{\Gamma} \frac{g}{h} \left(\rho_a' - \lambda + \frac{\partial \ln g}{\partial \ln \xi} \right) - \varkappa \frac{g}{h} \left(\rho_a' + \frac{\partial \ln h}{\partial \ln \xi} \right)$$

In particular, for a perfect gas with constant specific heats, $\Gamma = \gamma$ while $\varkappa = \frac{1}{\gamma - 1}$.

4. 6. Eulerian Space Profiles.

As pointed out in paragraph 4.2 with reference to Fig. 4.1, the Eulerian space profiles are obtained by seeking a solution along $\tau = 1$ for $\phi = x$.

It follows therefore that, for this

purpose,

$$\frac{\partial \ln x}{\partial \ln \phi} = 1 \quad \text{while} \quad \frac{\partial \ln \tau}{\partial \ln \phi} = \frac{\partial \ln x}{\partial \ln \xi} = \frac{\partial \ln \tau}{\partial \ln \xi} = 0 \ . \tag{4.27}$$

The transformed dependent variables of Eq. (4.16) are reduced then to

$$\left\{ \begin{array}{l} M_k = A_k x \\[2ex] N_k = A_k x - f B_k \\[2ex] K_k = C_k x \end{array} \right. \tag{4.28}$$

and Eq. (4.16) becomes

$$\left(\varrho_a' + j + 1 - \frac{\lambda k}{2} \right) M_k + \frac{\partial M_k}{\partial \ln \xi} - \frac{\partial N_k}{\partial \ln x} = K_k \tag{4.29}$$

where explicity, on the basis of the matrix table, Eqs. (4.18), one obtains the following expressions :

k	$\dfrac{M_k}{hx^{j+1}}$	$\dfrac{N_k}{hx^{j+1}}$	$\dfrac{K_k}{hx^{j+1}}$
0	1	$1 - \dfrac{f}{x}$	ω_M
1	f	$f\left(1 - \dfrac{f}{x}\right) - \dfrac{g}{xh}$	$\dfrac{jg}{xh} + \omega_F$
2	$\sigma + \dfrac{f^2}{2}$	$\left(\sigma + \dfrac{f^2}{2}\right)\left(1 - \dfrac{f}{x}\right) - \dfrac{gf}{xh}$	ω_E

$$(4.30)$$

Thus,for the conservation of mass,with $k = 0$ Eq. (4.29)
yields

$$(\varrho'_a + j + 1) h x^{j+1} + \frac{\partial h x^{j+1}}{\partial \ln \xi} - \frac{\partial}{\partial \ln x} \left[h x^j (x - f) \right] = h x^{j+1} \omega_M$$

(4.31)

for the conservation of momentum, with $k = 1$, it gives

$$\left(\varrho'_a + j + 1 - \frac{\lambda}{2} \right) h x^{j+1} f + \frac{\partial}{\partial \ln \xi} \left[h x^{j+1} f \right] - \frac{\partial}{\partial \ln x} \left[h x^j f (x - f) - x^j g \right] = h x^{j+1} \left(\frac{jg}{xh} + \omega_F \right) \ .$$

(4.32)

and for the conservation of energy, with $k = 2$, it
becomes :

$$(\varrho'_a + j + 1 - \lambda) h x^{j+1} \left(\sigma + \frac{f^2}{2} \right) + \frac{\partial}{\partial \ln \xi} \left[h x^{j+1} \left(\sigma + \frac{f^2}{2} \right) \right] - \frac{\partial}{\partial \ln x} \left[h x^j \left(\sigma + \frac{f^2}{2} \right) (x - f) - x^j gf \right] = h x^{j+1} \omega_E$$

(4.33)

Expanding Eq. (4.31) and dividing by
$h x^{j+1}$ one gets :

$$(4.34) \quad \varrho'_a + \frac{\partial \ln f}{\partial \ln \xi} - \left(1 - \frac{f}{x} \right) \frac{\partial \ln f}{\partial \ln x} + \frac{f}{x} \left(\frac{\partial \ln f}{\partial \ln x} + j \right) = \omega_M$$

Subtracting Eq. (4.31) multiplied by f from Eq.
(4.32) one obtains :

$$- \frac{\lambda}{2} + \frac{\partial \ln f}{\partial \ln \xi} - \left(1 - \frac{f}{x} \right) \frac{\partial \ln f}{\partial \ln x} + \frac{g}{hxf} \frac{\partial \ln g}{\partial \ln x} = \frac{1}{f} (\omega_F - f \omega_M)$$

(4.35)

Subtracting, in turn, Eq. (4.31) multiplied by
$\left(\sigma + \dfrac{g}{h} - \dfrac{f^2}{2}\right)$ and Eq. (4.35) multiplied by $f^3 h x^{j+1}$
from Eq. (4.33) one arrives at :

$$-\lambda\sigma + \frac{\partial\sigma}{\partial\ln\xi} - \left(1 - \frac{f}{x}\right)\frac{\partial\sigma}{\partial\ln x} - \frac{g}{h}\left[\varrho_a' + \frac{\partial\ln h}{\partial\ln\xi} - \left(1 - \frac{f}{x}\right)\frac{\partial\ln h}{\partial\ln x}\right] =$$

$$= \omega_E - f\omega_F - \left(\sigma + \frac{g}{h} - \frac{f^2}{2}\right)\omega_M \ . \qquad (4.36)$$

Using Eqs. (4.26) to eliminate σ , the above can be
also written as follows :

$$\varrho_a' - \lambda + \frac{\partial\ln g}{\partial\ln\xi} - \left(1 - \frac{f}{x}\right)\frac{\partial\ln g}{\partial\ln x} - \Gamma\left[\varrho_a' + \frac{\partial\ln h}{\partial\ln\xi} - \left(1 - \frac{f}{x}\right)\frac{\partial\ln h}{\partial\ln x}\right] =$$

$$= \frac{\Gamma}{x+1}\frac{h}{g}\left[\omega_E - f\omega_F - \left(\sigma + \frac{g}{h} - \frac{f^2}{2}\right)\omega_M\right] \ . \quad (4.37)$$

The expanded non-dimensional form of
blast wave equations represented by Eqs. (4.33),(4.34)
and (4.36) has attained most popularity in scientific
literature as exemplifield by the texts of Sedov (6),
Korobeinikov et al (9), and Sakurai (10).

The blast wave equations can be expressed in a more concise, autonomous, form by the
introduction of two reduced variables :

$$F \equiv \frac{\tau}{x} f \qquad \text{and} \qquad Z \equiv \left(\frac{\tau}{x}\right)^2\frac{g}{h} \qquad (4.38)$$

representing in effect, but by no means explicitly, the local velocity of flow and the local velocity of sound, respectively.

Since in the present case $\tau = 1$, it follows that

$$(4.48a) \qquad F = \frac{f}{x} \qquad \text{and} \qquad Z = \frac{1}{x^2}\frac{g}{h}$$

With the use of these variables, Eqs. (4.33), and (4.36) can be written as follows :

$$(4.39) \quad \rho'_a + \frac{\partial \ln h}{\partial \ln \xi} - (1-F)\frac{\partial \ln h}{\partial \ln x} + F\left(\frac{\partial \ln F}{\partial \ln x} + j + 1\right) = \Phi_M$$

$$(4.40) \quad -\frac{\lambda}{2} + \frac{\partial \ln f}{\partial \ln \xi} - (1-F)\left[\frac{\partial \ln F}{\partial \ln x} + 1\right] + \frac{Z}{F}\frac{\partial \ln g}{\partial \ln x} = \Phi_F$$

$$\rho'_a - \lambda + \frac{\partial \ln g}{\partial \ln \xi} - (1-F)\frac{\partial \ln g}{\partial \ln x} - \Gamma\left[\rho'_a + \frac{\partial \ln h}{\partial \ln \xi} - (1-F)\frac{\partial \ln h}{\partial \ln x}\right] = \Phi_E$$
(4.41)

where

$$(4.42) \quad \begin{aligned} \Phi_M &\equiv \omega_M \\[1ex] \Phi_F &\equiv \frac{1}{f}\left(\omega_F - f\omega_M\right) \\[1ex] \Phi_E &\equiv \frac{\Gamma}{x+1}\frac{h}{g}\left[\omega_E - f\omega_F - \left(\sigma + \frac{g}{h} - \frac{f^2}{2}\right)\omega_M\right] \end{aligned} \Biggr\}$$

Equations (4.39), (4.40), (4.41) can be considered as a set of three algebraic equations

for three unknowns that represent the logarithmic
gradients with respect to x of F, h and g, respectively. Solving them algebraically for these gradients,
one obtains the following autonomous form of the blast
wave equations :

$$\underline{D}\frac{\partial \ln F}{\partial \ln x} = \underline{F}_x + \underline{F}_\xi + \underline{F}_\omega \tag{4.43}$$

$$\underline{D}\frac{\partial \ln h}{\partial \ln x} = \underline{H}_x + \underline{H}_\xi + \underline{H}_\omega \tag{4.44}$$

$$\underline{D}\frac{\partial \ln g}{\partial \ln x} = \underline{G}_x + \underline{G}_\xi + \underline{G}_\omega \tag{4.45}$$

As a consequence of the definition of Z, one has
also from the above :

$$\underline{D}\frac{\partial \ln Z}{\partial \ln x} = \underline{Z}_x + \underline{Z}_\xi + \underline{Z}_\omega \tag{4.46}$$

where

$$\begin{cases} \underline{D} \equiv (1-F)^2 - \Gamma z \\[2mm] \underline{F}_x \equiv (1-F)\left(F - \frac{\lambda+2}{2}\right) + \frac{Z}{F}\left[(j+1)\Gamma F - \lambda + \varrho'_a\right] \\[2mm] \underline{H}_x \equiv \frac{1}{1-F}\left[F\underline{F}_x + \{(j+1)F + \varrho'_a\}\underline{D}\right] \\[2mm] \underline{G}_x \equiv \frac{1}{1-F}\left[\Gamma F\underline{F}_x + \{(j+1)\Gamma F - \lambda + \varrho'_a\}\underline{D}\right] \\[2mm] \underline{Z}_x \equiv \underline{G}_x - \underline{H}_x - 2\underline{D} = \frac{\Gamma-1}{1-F}\left[F\underline{F}_x + \{(j+1)F - \frac{\lambda}{\Gamma-1}\}\underline{D}\right] - 2\underline{D} \end{cases} \tag{4.47}$$

and

$$\underline{F}_\xi = (1-F)\frac{\partial \ln f}{\partial \ln \xi} + \frac{Z^2}{F}\frac{\partial \ln g}{\partial \ln \xi}$$

$$\underline{H}_\xi = \frac{1}{1-F}\left[F\underline{F}_\xi + \frac{\partial \ln h}{\partial \ln \xi}\underline{D}\right]$$

(4.48)

$$\underline{G}_\xi = \frac{1}{1-F}\left[\Gamma F\underline{F}_\xi + \frac{\partial \ln g}{\partial \ln \xi}\underline{D}\right]$$

$$\underline{Z}'_\xi = \underline{G}_\xi - \underline{H}_\xi$$

while

$$\underline{F}_\omega = -(1-F)\Phi_F - \frac{Z^2}{F}(\Gamma\Phi_M + \Phi_E)$$

$$\underline{H}_\omega = \frac{1}{1-F}\left[F\underline{F}_\omega - \Phi_M\underline{D}\right]$$

(4.49)

$$\underline{G}_\omega = \frac{1}{1-F}\left[\Gamma F\underline{F}_\omega - (\Gamma\Phi_M + \Phi_E)\underline{D}\right]$$

$$\underline{Z}_\omega = \underline{G}_\omega - \underline{H}_\omega$$

For a sourceless flow all the functions with subscript ω vanish. For self- similar solutions all the functions with subscript ξ are annihilated, while Γ and ϱ'_a can be taken into account only as constants, the latter limiting the variation of the atmospheric density into which the wave propagates to a power dependence on the distance from the center of explosion. Blast wave equations in the autonomous form for self-similar flows, corresponding to the case of $\varrho'_a = 0$

(i.e. ϱ_a = const.), were presented by Courant and Friedrichs (11), under the name of equations for "progressing waves", while algebraic solutions for the case of $\varrho'_a = \lambda - (j+1) = $ const were given by Sedov (6).

4.7. Eulerian Time Profiles.

Referring again to Fig. 4.1, the Eulerian time profiles are obtained by seeking a solution along $x = 1$ for $\Phi = \tau$. It follows then that

$$\frac{\partial \ln \tau}{\partial \ln \Phi} = 1 \quad \text{while} \quad \frac{\partial \ln x}{\partial \ln \Phi} = \frac{\partial \ln x}{\partial \ln \xi} = \frac{\partial \ln \tau}{\partial \ln \xi} = 0 \tag{4.50}$$

The generalized dependent variables are therefore reduced to

$$\begin{cases} M_k = -f B_k \tau \mu \\[2mm] N_k = A_k - f B_k \tau \\[2mm] K_k = -C_k \tau \mu \end{cases} \tag{4.51}$$

and Eqs. (4.16) and (4.18) with the use of the relation between μ and λ given by Eq. (4.13), yield

$$(4.52) \quad \left[\varrho'_a + j - \frac{\lambda}{2}(k+1) + \frac{1}{\mu} \right] \frac{M_k}{\mu} + \frac{\partial}{\partial \ln \xi}\left[\frac{M_k}{\mu}\right] - \frac{1}{\mu}\frac{\partial N_k}{\partial \ln \tau} = \frac{K_k}{\mu}$$

$$(4.52)$$

where, explicitly, the dependent variables are expressed in terms of the following matrix :

(4.53)

k	$-\dfrac{M_k}{\mu h \tau}$	$\dfrac{N_k}{h \tau}$	$-\dfrac{K_k}{\mu h \tau}$
0	f	$\dfrac{1-f\tau}{\tau}$	ω_M
1	$f^2 + \dfrac{g}{h}$	$f\left(\dfrac{1-f\tau}{\tau}\right) - \dfrac{g}{h}$	$j\dfrac{g}{h} + \omega_F$
2	$f\left(\sigma + \dfrac{f^2}{2}\right) + \dfrac{fg}{h}$	$\left(\sigma + \dfrac{f^2}{2}\right)\left(\dfrac{1-f\tau}{\tau}\right) - \dfrac{fg}{h}$	ω_E

Thus, for the conservation of mass Eq. (4.52) with $k = 0$ yields

$$\left(\varrho'_a + j - \frac{\lambda}{2} + \frac{1}{\mu}\right) h f \tau + \frac{\partial}{\partial \ln \xi}[hf\tau] + \frac{1}{\mu}\frac{\partial}{\partial \ln \tau}[h(1-f\tau)] = \tau h \omega_M$$

(4.54)

for the conservation of momentum, with $k = 1$, it gives

$$\left(\varrho'_a + j - \lambda + \frac{1}{\mu}\right)[\tau(hf^2 + g)] + \frac{\partial}{\partial \ln \xi}[\tau(hf^2 + g)] +$$

(4.55)
$$+ \frac{1}{\mu}\frac{\partial}{\partial \ln \tau}[hf(1-f\tau) - \tau g] = \tau[jg + h\omega_F]$$

and for the conservation of energy, with $k = 2$, it becomes

$$\left(\varrho'_a + j - \frac{3}{2}\lambda + \frac{1}{\mu}\right)\left[f\tau h\left(\sigma + \frac{f^2}{2} + \frac{g}{h}\right)\right] + \frac{\partial}{\partial \ln \xi}\left[hf\tau\left(\sigma + \frac{f^2}{2} + \frac{g}{h}\right)\right] +$$

(4.56)
$$+ \frac{1}{\mu}\frac{\partial}{\partial \ln \tau}\left[h\left(\sigma + \frac{f^2}{2}\right)(1-f\tau) - f\tau g\right] = h\omega_E \tau$$

Expanding Eq. (4.54) and dividing by $hf\tau$ one gets then

$$\left(\varphi_a' + \dot{\jmath} - \frac{\lambda}{2}\right) + \frac{\partial \ln h}{\partial \ln \xi} + \frac{1-f\tau}{\mu f \tau} \frac{\partial \ln h}{\partial \ln \tau} + \frac{\partial \ln f}{\partial \ln \xi} - \frac{1}{\mu} \frac{\partial \ln f}{\partial \ln \tau} = \frac{1}{f} \omega_M . \quad (4.57)$$

Subtracting Eq. (4.54) multiplied by f from Eq. (4.55) one obtains

$$-\frac{\lambda}{2} + \frac{\partial \ln f}{\partial \ln \xi} + \frac{1-f\tau}{\mu f \tau} \frac{\partial \ln f}{\partial \ln \tau} + \frac{g}{hf^2} \left[\varphi_a' - \lambda + \frac{\partial \ln g}{\partial \ln \xi} - \frac{1}{\mu} \frac{\partial \ln g}{\partial \ln \tau} \right] = \frac{1}{f^2} (\omega_F - f\omega_M) .$$
$$(4.58)$$

Subtracting, in turn, Eq. (4.54) multiplied by $\left(\sigma + \frac{g}{h} + \frac{f^2}{2}\right)$ and Eq. (4.58) multiplied by $\tau f^3 h$ from Eq. (4.56) one arrives at :

$$-\lambda \sigma + \frac{\partial \sigma}{\partial \ln \xi} + \frac{1-f\tau}{\mu f \tau} \frac{\partial \sigma}{\partial \ln \tau} - \frac{g}{h} \left[\varphi_a' + \frac{\partial \ln h}{\partial \ln \xi} + \frac{1-f\tau}{\mu f \tau} \frac{\partial \ln h}{\partial \ln \tau} \right] =$$
$$= \frac{1}{f} \left[\omega_E - f\omega_F - \left(\sigma + \frac{g}{h} - \frac{f^2}{2}\right) \right] . \quad (4.59)$$

If $e = e(p, \varphi)$, the above can be expressed also by the use of Eqs. (4.26) as follows :

$$\varphi_a' - \lambda + \frac{\partial \ln g}{\partial \ln \xi} + \frac{1-f\tau}{\mu f \tau} \frac{\partial \ln g}{\partial \ln \tau} - \Gamma \left[\varphi_a' + \frac{\partial \ln h}{\partial \ln \xi} + \frac{1-f\tau}{\mu f \tau} \frac{\partial \ln h}{\partial \ln \tau} \right] =$$
$$= \frac{\Gamma}{\chi + 1} \frac{h}{g} \frac{1}{f} \left[\omega_E - f\omega_F - \left(\sigma + \frac{g}{h} - \frac{f^2}{2}\right) \omega_M \right] . \quad (4.60)$$

To express the blast wave equations in the autonomous form one utilizes again the reduced variables, which in this case, since $x = 1$, are given by :

(4.38b) $\qquad F = f\tau \qquad$ and $\qquad Z = \tau^2 \dfrac{g}{h}$

In terms of these variables Eqs. (4.57), (4.58), and (4.60) become :

$$\left(\varrho'_a + j - \frac{\lambda}{2}\right) + \frac{\partial \ln h}{\partial \ln \xi} + \frac{1-F}{F\mu}\frac{\partial \ln h}{\partial \ln \tau} + \frac{\partial \ln f}{\partial \ln \xi} - \frac{1}{\mu}\left(\frac{\partial \ln F}{\partial \ln \tau} - 1\right) = \tilde{\Phi}_M$$

(4.61)

$$-\frac{\lambda}{2} + \frac{\partial \ln f}{\partial \ln \xi} + \frac{1-F}{F\mu}\left(\frac{\partial \ln F}{\partial \ln \tau} - 1\right) + \frac{Z}{f^2}\left[\varrho'_a - \lambda + \frac{\partial \ln g}{\partial \ln \xi} - \frac{1}{\mu}\frac{\partial \ln g}{\partial \ln \tau}\right] = \tilde{\Phi}_F$$

(4.62)
and

$$\varrho'_a - \lambda + \frac{\partial \ln g}{\partial \ln \xi} + \frac{1-F}{F\mu}\frac{\partial \ln g}{\partial \ln \tau} - \Gamma\left[\varrho'_a + \frac{\partial \ln h}{\partial \ln \xi} + \frac{1-F}{F\mu}\frac{\partial \ln h}{\partial \ln \tau}\right] = \tilde{\Phi}_E$$

(4.63)

where

$$\left.
\begin{aligned}
\tilde{\Phi}_M &= \frac{1}{f}\,\omega_M \\[2mm]
(4.64)\qquad \tilde{\Phi}_F &= \frac{1}{f^2}\left(\omega_F - f\omega_M\right) \\[2mm]
\tilde{\Phi}_E &= \frac{\Gamma}{X+1}\frac{h}{gf}\left[\omega_E - f\omega_F - \left(\sigma + \frac{g}{h} - \frac{f^2}{2}\right)\omega_M\right]
\end{aligned}
\right\}$$

Solving these equations algebraically, as before, for the logarithmic gradients of F, h and g with respect to τ , one obtains now :

$$\tilde{\underline{D}} \frac{1}{\mu} \frac{\partial \ln F}{\partial \ln \tau} = \tilde{F}_\tau + \tilde{F}_\xi + \tilde{F}_\omega \qquad (4.65)$$

$$\tilde{\underline{D}} \frac{1}{\mu} \frac{\partial \ln h}{\partial \ln \tau} = \tilde{H}_\tau + \tilde{H}_\xi + \tilde{H}_\omega \qquad (4.66)$$

$$\tilde{\underline{D}} \frac{1}{\mu} \frac{\partial \ln g}{\partial \ln \tau} = \tilde{G}_\tau + \tilde{G}_\xi + \tilde{G}_\omega \qquad (4.67)$$

and, from the definition of Z one gets from the above,

$$\tilde{\underline{D}} \frac{1}{\mu} \frac{\partial \ln Z}{\partial \ln \tau} = \tilde{Z}_\tau + \tilde{Z}_\xi + \tilde{Z}_\omega \qquad (4.68)$$

where

$$
\left\{
\begin{array}{l}
\tilde{\underline{D}} = (1 - F)^2 - \Gamma z \\[2mm]
\tilde{\underline{F}}_\tau = -(1-F)\left(F - \frac{\lambda+2}{2}\right) - \frac{Z}{F}\left[(j+1)\Gamma F - \lambda + \varrho'_a\right] \\[2mm]
\tilde{\underline{H}}_\tau = \frac{F}{1-F}\left[\tilde{F}_\tau - (\varrho'_a + j + 1)\tilde{\underline{D}}\right] \\[2mm]
\tilde{\underline{G}}_\tau = \frac{F}{1-F}\left[\Gamma\tilde{F}_\tau - \{(j+1)\Gamma F - \lambda + \varrho'_a\}\tilde{\underline{D}}\right] \\[2mm]
\tilde{\underline{Z}}_\tau = \tilde{G}_\tau - \tilde{H}_\tau + (\lambda + 2)\tilde{\underline{D}}
\end{array}
\right. \qquad (4.69)
$$

and

$$\tilde{\underline{F}}_\xi = -[F(1-F)+\Gamma Z]\frac{\partial \ln f}{\partial \ln \xi} - \frac{Z}{F}\frac{\partial \ln g}{\partial \ln \xi} + \mu'\tilde{\underline{D}}$$

$$\tilde{\underline{H}}_\xi = \frac{F}{1-F}\left[\tilde{\underline{F}}_\xi - \left(\mu' + \frac{\partial \ln f}{\partial \ln \xi} + \frac{\partial \ln h}{\partial \ln \xi}\right)\tilde{\underline{D}}\right]$$

(4.70)
$$\tilde{\underline{G}}_\xi = \frac{\Gamma F}{1-F}\left[\tilde{\underline{F}}_\xi - \left(\mu' + \frac{\partial \ln f}{\partial \ln \xi} + \frac{1}{\Gamma}\frac{\partial \ln g}{\partial \ln \xi}\right)\tilde{\underline{D}}\right]$$

$$\tilde{\underline{Z}}_\xi = \tilde{\underline{G}}_\xi - \tilde{\underline{H}}_\xi + 2\mu'\tilde{\underline{D}}$$

while

$$\tilde{\underline{F}}_\omega = F(1-F)\tilde{\Phi}_F + Z(\Gamma\tilde{\Phi}_M + \tilde{\Phi}_E)$$

$$\tilde{\underline{H}}_\omega = \frac{F}{1-F}[\tilde{\underline{F}}_\omega + \tilde{\Phi}_M\tilde{\underline{D}}]$$

(4.71)
$$\tilde{\underline{G}}_\omega = \frac{\Gamma F}{1-F}\left[\tilde{\underline{F}}_\omega + \left(\tilde{\Phi}_M + \frac{1}{\Gamma}\tilde{\Phi}_E\right)\tilde{\underline{D}}\right]$$

$$\tilde{\underline{Z}}_\omega = \tilde{\underline{G}}_\omega - \tilde{\underline{H}}_\omega$$

The three groups of terms above correspond to those introduced in the previous case by Eqs. (4.47), (4.48) and (4.49).

4. 8. Lagrangian Time Profiles.

Again with reference to Fig. 4.1, time profiles in Lagrangian coordinates are obtained by seeking a solution with respect to $\phi = \tau$ under the restriction imposed by the condition $\left(\frac{\partial r}{\partial t}\right)_\xi = u$, or, in non-dimensional form,

$$\left(\frac{\partial x}{\partial \tau}\right)_{\xi} = \frac{w_n t_n}{r_n} f = \mu f \qquad (4.72)$$

consequently

$$\frac{\partial \ln \tau}{\partial \ln \phi} = 1 \;\; , \;\; \frac{\partial \ln \tau}{\partial \ln \xi} = 0 \quad \text{and} \quad \frac{\partial \ln x}{\partial \ln \phi} = \mu \frac{fx}{\tau} \quad (4.73)$$

Hence

$$\left\{ \begin{array}{l} M_k = \mu f \tau (A_k - B_k) \\[2mm] N_k = A_k x \zeta + f \tau (A_k - B_k) \\[2mm] K_k = -\mu C_k x \tau \zeta \end{array} \right. \qquad (4.74)$$

where

$$\zeta \equiv 1 - \frac{\tau}{x} f + \frac{\partial \ln x}{\partial \ln \xi} \qquad (4.75)$$

As before, with the use of Eq. (4.13), Eqs. (4.16) and (4.18) yield

$$\left[\rho_a' + j - \frac{\lambda}{2}(k+1) + \frac{1}{\mu} \right] \frac{M_k}{\mu} + \frac{\partial}{\partial \ln \xi}\left(\frac{M_k}{\mu}\right) - \frac{1}{\mu}\frac{\partial N_k}{\partial \ln \tau} = \frac{K_k}{\mu}$$

$$(4.76)$$

where, explicitly, the dependent variables are given by the following matrix :

(4.77)

k	$-\dfrac{M_k}{\mu h x^{i+1}\zeta\tau}$	$\dfrac{N_k}{h x^{i+1}\zeta\tau}$	$-\dfrac{K_k}{\mu h x^{i+1}\zeta\tau}$
0	0	$\dfrac{1}{\tau}$	ω_M
1	$\dfrac{g}{x\zeta h}$	$\dfrac{f}{\tau}-\dfrac{g}{x\zeta h}$	$\dfrac{ig}{xh}+\omega_F$
2	$\dfrac{fg}{x\zeta h}$	$\dfrac{\sigma+f^2/2}{\tau}-\dfrac{fg}{x\zeta h}$	ω_E

Thus, for the conservation of mass, Eq. (4.76) with k = 0, yields

$$(4.78)\qquad \frac{1}{\mu}\frac{\partial}{\partial\ln\tau}\left[hx^{i+1}\zeta\right]=\tau h x^{i+1}\zeta\omega_M$$

for the conservation of momentum, with k = 1, it gives

$$\left(\varrho'_a+i-\lambda+\frac{1}{\mu}\right)x^i\tau g+\frac{\partial}{\partial\ln\xi}\left(x^i g\tau\right)+\frac{1}{\mu}\frac{\partial}{\partial\ln\tau}\left[hx^{i+1}\zeta f-x^i\tau g\right]=$$

$$(4.79)\qquad\qquad = \tau h x^{i+1}\zeta\left(\frac{ig}{xh}+\omega_F\right)$$

and for the conservation of energy, with k = 2, it becomes

$$\left(\varrho'_a+i-\frac{3\lambda}{2}+\frac{1}{\mu}\right)x^i\tau fg+\frac{a}{\partial\ln\xi}\left(x^i\tau fg\right)+\frac{1}{\mu}\frac{\partial}{\partial\ln\tau}\left[hx^{i+1}\zeta\left(\sigma+\frac{f^2}{2}\right)-\right.$$

$$(4.80)\qquad\qquad \left.-x^i\tau fg\right]=\tau h x^{i+1}\zeta\omega_E$$

Integrating Eq. (4.78), subject to boundary condition $h_a x^{j+1} \zeta_a = 1$, it follows that :

$$h x^{j+1} \zeta = \exp\left[\mu \int_1^\tau \omega_M \, d\tau \right] = \Psi \qquad (4.81)$$

and, with the use of Eq. (4.72), one obtains then :

$$-\frac{\lambda}{2} + \frac{\partial \ln f}{\partial \ln \xi} - \frac{1}{\mu}\frac{\partial \ln f}{\partial \ln \tau} + \frac{x\zeta}{f\tau}\left[\frac{1}{\mu}\frac{\partial \ln h}{\partial \ln \tau} + j\frac{f\tau}{x} \right] = \frac{x\zeta}{f}\,\omega_M \;.$$
$$(4.82)$$

Subtracting Eq. (4.78) multiplied by f and Eq. (4.82) multiplied by $\dfrac{j\tau g}{xh}$ from Eq. (4.79) one gets :

$$\varrho_a' - \lambda + \frac{\partial \ln g}{\partial \ln \xi} - \frac{1}{\mu}\frac{\partial \ln g}{\partial \ln \tau} + \frac{h\zeta}{\mu g}\frac{\partial f}{\partial \tau} = x\zeta\frac{h}{g}\,(\omega_F - f\omega_M) \;.$$
$$(4.83)$$

Finally, subtracting Eq. (4.78) multiplied by $\left(\sigma + \dfrac{g}{h} + \dfrac{f^2}{2} \right)$ and Eq. (4.83) multiplied by $x^j g f \tau$ from Eq. (4.80), one arrives at :

$$\frac{\partial \sigma}{\partial \tau} - \frac{g}{h}\frac{\partial \ln h}{\partial \tau} = \mu \left[\omega_E - f\omega_F - \left(\sigma + \frac{g}{h} - \frac{f^2}{2} \right)\omega_M \right],$$
$$(4.84)$$

which, if $e = e(p, \varrho)$, can be expressed, by the use of Eq. (4.26), as

$$\frac{\partial \ln g}{\partial \ln \tau} - \Gamma\frac{\partial \ln h}{\partial \ln \tau} = \frac{\Gamma\mu}{\chi+1}\frac{h}{g}\left[\omega_E - f\omega_F - \left(\sigma + \frac{g}{h} - \frac{f^2}{2} \right)\omega_M \right].$$
$$(4.85)$$

The equations for the time profiles in Lagrangian co-

ordinates can be expressed in an autonomous form by
introducing again the reduced variables

(4.38) $F = \dfrac{\tau}{x} f$ and $Z = \left(\dfrac{\tau}{x}\right)^2 \dfrac{g}{h}$

Equations (4.82), (4.83) and (4.85),
with the help of Eq. (4.72), become then, respectively,

(4.86) $-\dfrac{\lambda}{2} + \dfrac{\partial \ln f}{\partial \ln \xi} - \dfrac{1}{\mu} \dfrac{\partial \ln F}{\partial \ln \tau} - F + \dfrac{1}{\mu} + \dfrac{1}{F}\left(1 - F + \dfrac{\partial \ln x}{\partial \ln \xi}\right)\left(\dfrac{1}{\mu}\dfrac{\partial \ln h}{\partial \ln \tau} + jF\right) = \hat{\Phi}_M$

(4.87) $\varrho'_a - \lambda + \dfrac{\partial \ln g}{\partial \ln \xi} - \dfrac{1}{\mu}\dfrac{\partial \ln g}{\partial \ln \tau} + \dfrac{F}{Z}\left(1 - F + \dfrac{\partial \ln x}{\partial \ln \xi}\right)\left(\dfrac{1}{\mu}\dfrac{\partial \ln F}{\partial \ln \tau} + F - \dfrac{1}{\mu}\right) = \hat{\Phi}_F$

(4.88) $\dfrac{1}{\mu}\dfrac{\partial \ln g}{\partial \ln \tau} - \dfrac{\Gamma}{\mu}\dfrac{\partial \ln h}{\partial \ln \tau} = \hat{\Phi}_E$

where

(4.89)
$$\hat{\Phi}_M \equiv \dfrac{x\zeta}{f}\,\omega_F$$
$$\hat{\Phi}_F \equiv \dfrac{x\zeta h}{g}\,(\omega_F - f\omega_M)$$
$$\hat{\Phi}_E \equiv \dfrac{\Gamma}{x+1}\dfrac{h}{g}\left[\omega_E - f\omega_F - \left(\sigma + \dfrac{g}{h} - \dfrac{f^2}{2}\right)\omega_M\right]$$

Solving, as before, the above equations algebraically
for the logarithmic gradients with respect to τ , one
obtains now the autonomous form of the blast wave equa-

tions as follows :

$$\hat{\underline{D}}\,\frac{1}{\mu}\frac{\partial \ln F}{\partial \ln \tau} = \hat{\underline{F}}_\tau + \hat{\underline{F}}_\xi + \hat{\underline{F}}_\omega \qquad (4.90)$$

$$\hat{\underline{D}}\,\frac{1}{\mu}\frac{\partial \ln h}{\partial \ln \tau} = \hat{\underline{H}}_\tau + \hat{\underline{H}}_\xi + \hat{\underline{H}}_\omega \qquad (4.91)$$

$$\hat{\underline{D}}\,\frac{1}{\mu}\frac{\partial \ln g}{\partial \ln \tau} = \hat{\underline{G}}_\tau + \hat{\underline{G}}_\xi + \hat{\underline{G}}_\omega \qquad (4.92)$$

whence, according to the definition of Z,

$$\hat{\underline{D}}\,\frac{1}{\mu}\frac{\partial \ln Z}{\partial \ln \tau} = \hat{\underline{Z}}_\tau + \hat{\underline{Z}}_\xi + \hat{\underline{Z}}_\omega \qquad (4.93)$$

where

$$
\left\{
\begin{aligned}
\hat{\underline{D}} &\equiv \left(1 - F + \frac{\partial \ln x}{\partial \ln \xi}\right)^2 - \Gamma Z \\[2mm]
\hat{\underline{F}}_\tau &\equiv -\left(1 - F + \frac{\partial \ln x}{\partial \ln \xi}\right)\left[\left(1 - F + \frac{\partial \ln x}{\partial \ln \xi}\right)\left(F - \frac{\lambda+2}{2}\right) + \frac{Z}{F}\left\{(j+1)\Gamma F - \lambda + \varrho'_a\right\}\right] \\[2mm]
\hat{\underline{H}}_\tau &\equiv \frac{F}{1 - F + \partial \ln x/\partial \ln \xi}\,\hat{\underline{F}}_\tau - (j+1)F\hat{\underline{D}} \qquad (4.94) \\[2mm]
\hat{\underline{G}}_\tau &\equiv \Gamma \hat{\underline{H}}_\tau \\[2mm]
\hat{\underline{Z}}_\tau &\equiv (\Gamma - 1)\hat{\underline{H}}_\tau - \left(F - \frac{\lambda+2}{2}\right)\hat{\underline{D}}
\end{aligned}
\right.
$$

and

$$
\left\{
\begin{aligned}
\hat{\underline{F}}_\xi &\equiv -\left[\left(\frac{\partial \ln f}{\partial \ln \xi} - \frac{\partial \ln x}{\partial \ln \xi}\right)\Gamma F + \left(1 - F + \frac{\partial \ln x}{\partial \ln \xi}\right)\frac{\partial \ln g}{\partial \ln \xi}\right]\frac{Z}{F} + \mu'\hat{\underline{D}} \\[2mm]
\hat{\underline{H}}_\xi &\equiv \frac{F}{1 - F + \partial \ln x/\partial \ln \xi}\left[\hat{\underline{F}}_\xi - \left(\frac{\partial \ln f}{\partial \ln \xi} - \frac{\partial \ln x}{\partial \ln \xi} + \mu'\right)\hat{\underline{D}}\right]
\end{aligned}
\right.
$$
$$(4.95a)$$

$$\hat{G}_\xi = \Gamma \hat{H}_\xi$$

(4.95b)

$$\hat{Z}_\xi = (\Gamma - 1)\hat{H}_\xi + 2\mu \hat{D}$$

while

$$\hat{F}_\omega = \left[\Gamma F \hat{\Phi}_M + \left(1 - F + \frac{\partial \ln x}{\partial \ln \xi}\right)(\hat{\Phi}_F + \hat{\Phi}_E)\right]\frac{Z}{F}$$

$$\hat{H}_\omega = \frac{F(\hat{F}_\omega + \hat{\Phi}_M \hat{D})}{1 - F + \partial \ln x/\partial \ln \xi}$$

(4.96)

$$\hat{G}_\omega = \Gamma \hat{H}_\omega + \hat{\Phi}_E \hat{D}$$

$$\hat{Z}_\omega = (\Gamma - 1)\hat{H}_\omega + \hat{\Phi}_E \hat{D} .$$

The above grouping of terms is again the same as it was in the previous cases.

4.9. Boundary Conditions and Integral Relations.

The flow field of a blast wave is, by definition, bounded by a gasdynamic discontinuity. Thus, at the front, that is at $x = \tau = 1$ one has, on the basis of the Hugoniot relations, § 3.6, the following boundary conditions :

(4.97)
$$h_n = \nu^{-1} = \frac{P + \beta_P}{(P_G + \beta_P) + \beta_\gamma (P - P_G)}$$

$$f_n = F_n = \frac{U_i}{M_n} = 1 - \nu = (1 - \beta_\nu)\frac{P - P_G}{P + \beta_P} \qquad (4.98)$$

and

$$g_n = \frac{1}{\Gamma_a M_n^2}\frac{P_n}{P_a} = \frac{(1-\nu)P}{P-1} = (1-\beta_\nu)\frac{P - P_G}{P + \beta_P}\frac{P}{P-1} \qquad (4.99)$$

while

$$Z_n = \frac{g_n}{h_n} = \frac{1-\nu}{P-1}P\nu = (1-\beta_\nu)[(P_G+\beta_\nu)+\beta_\nu(P-P_G)]\frac{P - P_G}{(P+\beta_P)^2}\frac{P}{P-1}$$
$$(4.100)$$

In all the above equations P can be expressed in terms of M_n by virtue of Eq. (3.46).

For a shock front producing a change of state expressible in terms of a Rankine-Hugoniot hyperbola for which $P_G = 1$, one has a straightforward relation between P and M_n, Eq. (3.52), so that, with its use, Eqs. (4.79) – (4.100) yield

$$h_n = \frac{M_n^2}{1 + \beta_\nu(M_n^2 - 1)} = \frac{1}{\beta_\nu + (1 - \beta_\nu)y} \qquad (4.101)$$

$$f_n = F_n = (1 - \beta_\nu)(1 - \frac{1}{M^2}) = (1 - \beta_\nu)(1 - y) \qquad (4.102)$$

$$(4.103) \quad g_n = (1-\beta_v)\left(1 - \frac{\beta_p}{1+\beta_p}\frac{1}{M_n^2}\right) = \frac{1-\beta_v}{1+\beta_p}(1 +\beta_p - \beta_p y)$$

$$Z_n = (1-\beta_v)\beta_v\left(1 - \frac{\beta_p}{1+\beta_p}\frac{1}{M_n^2}\right)\left(1 + \frac{\beta_v}{1-\beta_v}\frac{1}{M_n^2}\right) = \frac{1-\beta_v}{1+\beta_p}(1+\beta_p-\beta_p y)[\beta_v+(1-\beta_v)y]$$

(4.104)

For a perfect gas with γ = const.

$$\beta_p = \beta_v = \frac{\gamma-1}{\gamma+1} \quad \text{and Eqs. (4.101) - (4.104) become}$$

$$(4.105) \qquad\qquad h_n = \frac{\gamma+1}{\gamma-1+2y}$$

$$(4.106) \qquad\qquad f_n = F_n = \frac{2}{\gamma+1}(1-y)$$

$$(4.107) \qquad\qquad g_n = \frac{\gamma-1}{\gamma(\gamma+1)}\left(\frac{2\gamma}{\gamma-1} - y\right)$$

$$(4.108) \qquad\qquad Z_n = \frac{2(\gamma-1)}{\gamma(\gamma+1)^2}\left(\frac{2\gamma}{\gamma-1} - y\right)\left(\frac{\gamma-1}{2}+y\right)$$

At the inner boundary the blast wave has usually the condition of rest at the point of symmetry, i.e.,

$$(4.109) \qquad\qquad f(0,\tau) = 0$$

For a blast wave generated by a piston starting at the center of symmetry, instead of the above, one has the condition that a Φ = const line must coincide with a particle path i.e.

$$(4.110a) \qquad\qquad u_p = \left(\frac{\partial r_p}{\partial t}\right)_\Phi$$

If at a $\Phi = $ const line $x = \tau = $ const., as it is, indeed, the case in a self-similar flow field, §4.9, then

$$\left(\frac{\partial r_n}{\partial t}\right)_\Phi = \left[\frac{\partial(x_p r_n)}{\partial(\tau_p t)}\right]_{\Phi_p = x_p = \tau_p} = \frac{x_p}{\tau_p}\frac{dr_n}{dt} \qquad (4.110b)$$

so that, at the piston face,

$$F_p \equiv \frac{\tau_p w_p}{x_p w_n} = 1 \quad .$$

The integral relations for a blast wave are the expressions of the principles of global mass and energy conservations. The first requires that

$$\int_0^{r_n} \rho\, r^j\, dr = \frac{\rho_a\, r_n^{j+1}}{j+1+\rho_a'}\left(1 + \int_0^t \Omega_M\, dt\right) \qquad (4.111)$$

or, in a non-dimensional form,

$$\int_0^1 h\, x^j\, dx = \frac{1}{j+1+\rho_a'}\left(1 + \mu \int_0^\tau \bar\omega_m\, d\tau\right) \qquad (4.112)$$

where $\bar\Omega_M$ and $\bar\omega_M$ express the space averages of the fractional mass addition rates. The second can be expressed in terms of the energy integral

$$E_j \equiv \int_0^{r_n}\left(e - e_a + \frac{u^2}{2}\right)\rho\, r^j\, dr \qquad (4.113)$$

where

$$E_j = \begin{cases} \quad \text{blast wave energy per unit area for } j = 0 \\ \dfrac{1}{2\pi} \text{ blast wave energy per unit length for } j = 1 \\ \dfrac{1}{4\pi} \text{ blast wave energy for } j = 2 \end{cases}$$

the integration being taken along the abscissa $t = t_n$.
If for example, the blast wave is piston driven,

$$(4.14) \qquad\qquad E_j = \int_0^{r_p} p\, r^j\, dr$$

the integration being taken along the piston path in
the time-space domain (piston world line).

The most common case is that of a
decaying blast wave initiated by the deposition of a
finite amount of energy in a compressible medium and
propagating without any other energy exchange with
the surroundings. For such a constant energy case, E_j
can be expressed in terms of a reference radius

$$(4.15) \qquad\qquad r_0 \equiv \left(\frac{E_j}{p_a}\right)^{\frac{1}{j+1}}$$

and, taking account of Eq. (4.112), with $\bar{\omega}_M = 0$ Eq. (4.113) can
be reduced to the following non-dimensional form

$$(4.16) \qquad \frac{y}{\Gamma_a \xi^{j+1}} + \frac{\sigma_a}{j+1+Q_0'} - \int_0^1 \left(\sigma + \frac{f^2}{2}\right) h x^j\, dx \ .$$

For a perfect gas with $\gamma = \Gamma_a = \text{const}$, $\sigma_a = \dfrac{1}{\gamma - 1}\dfrac{g}{h}$

while $\qquad \sigma_a = \dfrac{1}{\gamma-1}\,\dfrac{y}{\gamma}\qquad$ and the above becomes

$$\left[\frac{1}{\xi^{j+1}}+\frac{1}{(\gamma-1)(j+1+\varphi_a')}\right]\frac{y}{\gamma} = \int_0^1\left(\frac{g}{\gamma-1}+\frac{hf^2}{2}\right)x^j dx$$

$$= \int_0^1\left(\frac{Z}{\gamma-1}+\frac{F^2}{2}\right)hx^{j+2}dx\,.$$

$$(4.117)$$

4. 10. Self-Similar Flow Fields

In order to illustrate a solution of blast wave equations, the simplest case of a self-similar flow field with a constant velocity modulus μ, is presented here. Included, however, in the analysis is the possibility of density variation in the atmosphere into which the wave propagates, provided that it can be expressed in terms of a power law relation with the distance from the center of explosion, since taking into account such a variation does not introduce any complication.

As a consequence of the assumption that μ = const., (Eq. 4.13) becomes simply

$$\mu = \frac{2}{\lambda+2}\qquad(4.118)$$

and all the three sets of autonomous equations can be reduced immediately to a single form by the introduction of a non-dimensional self-similarity variable :

$$\varphi = \frac{x}{\tau^\mu}\qquad(4.119)$$

Thus,

$$(4.120) \quad F \equiv \frac{\tau}{x} f = \frac{\tau^{1-\mu}}{\varphi} f \quad \text{and} \quad Z \equiv \left(\frac{\tau}{x}\right)^2 \frac{g}{h} = \left(\frac{\tau^{1-\mu}}{\varphi}\right)^2 \frac{g}{h}$$

Moreover, from the definition of φ it follows that, in the case of space profiles in Eulerian coordinates, since then $\tau = 1$, one has, for any function F

$$(4.121) \qquad\qquad \frac{dF}{d\ln x} = \frac{dF}{d\ln\varphi}$$

for time profiles in Eulerian coordinates, since then $x = 1$, one obtains

$$(4.122) \qquad\qquad \frac{1}{\mu}\frac{dF}{d\ln\tau} = -\frac{dF}{d\ln\varphi}$$

while for time profiles in Lagrangian coordinates, as a consequence of the kinematic relation, Eq. (4.72), one gets

$$(4.123) \qquad\qquad \frac{1}{\mu}\frac{dF}{d\ln\tau} = -(1-F)\frac{dF}{d\ln\varphi}$$

With the use of the above transformations, Eqs.(4.43), (4.65) and (4.90) are all reduced to a single form :

$$D\frac{d\ln F}{d\ln\varphi} = (1-F)\left(F-\frac{\lambda+2}{2}\right)F + \left[(j+1)\Gamma F - (\lambda - \varrho_a')\right]\frac{Z}{F}$$

$$(4.124)$$

while Eqs. (4.46), (4.68) and (4.93) are similarly

reduced to :

$$(4.125)$$

$$\underline{D}\frac{d\ln Z}{d\ln\varphi} = (\Gamma - 1)\left(F - \frac{\lambda+2}{2}\right)F + \frac{\Gamma-1}{1-F}[(j+1)\Gamma F - (\lambda - \varrho_a')]Z + \frac{\lambda+2}{1-F}\left(\frac{F}{\gamma} - 1\right)\underline{D}.$$

In the above, as in the general cases of Eulerian

profiles,

$$\underline{D} \equiv (1 - F)^2 - \Gamma Z \qquad\qquad (4.126)$$

while

$$\gamma \equiv \frac{\lambda+2}{(\Gamma-1)(j+1)+2} \qquad\qquad (4.127)$$

and, for the sake of self-similarity, one has to

have an equation of state of such a form that $\Gamma \equiv \Gamma(Z)$.

Equations (4.124) and (4.125) can be

solved in terms of F which is for this purpose adop-

ted as a parameter. The solution is, in effect, go-

verned by a single differential equation obtained

immediately from the ratio between these two equations,

namely :

$$\frac{d\ln Z}{d\ln F} = \frac{F}{1-F}\left[(\Gamma+1) + \frac{(\lambda+2)\left(\frac{F}{\gamma}-1\right)\underline{D}}{(1-F)\left(F-\frac{\lambda+2}{2}\right)F + [(j+1)\Gamma F - (\lambda - \varrho_a')]Z}\right] \cdot (4.128)$$

The coordinates F and Z define thus a phase plane

where any specific solution is represented by an in-

tegral curve passing through a point (F_n, Z_n) fixed by
the boundary conditions associated with the gasdynamic
discontinuity at the front. With $Z = Z(F)$ thus determined, Eq. (4.124) gives, by quadrature, $\varphi = \varphi(F)$. At
the same time all the remaining equations can be
integrated immediately, yielding the so-called adiabatic integral. In order to derive the latter, it is
convenient to introduce a new parameter

$$(4.129) \qquad\qquad \alpha \equiv \frac{\lambda + (\Gamma - 1)\varrho_a'}{j + 1 + \varrho_a'}$$

Muttiplying then by α Eq.(4.39) and adding to Eq.
(4.41), with proper account taken of the definition
of   Eq. (4.120), one obtains :

$$(4.130) \qquad \frac{d\ln g}{d\ln\varphi} - \Gamma\frac{d\ln h}{d\ln\varphi} = -\alpha\,\frac{d\ln[hx^{j+1}(1 - F)]}{d\ln\varphi}$$

An identical result can be also obtained from Eq.
(4.60) and (4.61) by noting that in this case $x = 1$,
while in the case of Lagrangian coordinates one has,
for a sourceless, self-similar flow field, from Eq.
(4.81),

$$(4.131) \quad hx^{j+1}\zeta = hx^{j+1}(1 - F) = h(\varphi\tau^\mu)^{j+1}(1 - F) = 1$$

so that, in accordance with Eq. (4.85), the right hand

side of Eq. (4.130) is zero.

If Γ = const , the adiabatic integral for all three cases can be expressed therefore, on the basis of Eq. (4.130) as

$$\left(\frac{g}{g_n}\right)\left(\frac{h}{h_n}\right)^{-\Gamma} = \left[\frac{h}{h_n}\,x^{j+1}\left(\frac{1-F}{1-F_n}\right)\right]^{-\alpha} \qquad (4.132)$$

where, for Eulerian time profiles, $x = 1$, while for time profiles in Lagrangian coordinates the right hand side is unity.

Now then, profiles in any system of coordinates are given, respectively, by φ , according to its definition introduced at the outset of this section, the expressions for F and Z of Eqs. (4.120), and the adiabatic integral, Eq. (4.132).

Thus, explicity, for Eulerian space profiles one has

$$\frac{r}{r_n} = x = \varphi \qquad (4.133)$$

$$\frac{u}{u_n} = \frac{f}{f_n} = \frac{F}{F_n}\,x \qquad (4.134)$$

$$\frac{p}{p_n} = \frac{g}{g_n} = \left(\frac{Z}{Z_n}\right)^2\left(\frac{h}{h_n}\right)x^2 \qquad (4.135)$$

$$(4.136) \qquad \frac{\varrho}{\varrho_n} = \frac{h}{h_n} = \left[\left(\frac{Z}{Z_n}\right)^2 \left(\frac{1-F}{1-F_n}\right)^\alpha x^{(i+1)\alpha+2} \right]^{\frac{1}{\Gamma-1-\alpha}}$$

$$(4.137) \qquad \frac{T}{T_n} = \left(\frac{g}{g_n}\right)\left(\frac{h}{h_n}\right)^{-1} = \left(\frac{Z}{Z_n}\right)^2 x^2$$

For Eulerian time profiles one gets

$$(4.138) \qquad \frac{t}{t_n} = \tau = \varphi^{-\frac{\lambda+2}{2}}$$

$$(4.139) \qquad \frac{u}{u_n} = \frac{f}{f_n} = \frac{F}{F_n}\tau^{-1}$$

$$(4.140) \qquad \frac{p}{p_n} = \frac{h}{h_n} = \left(\frac{Z}{Z_n}\right)^2\left(\frac{h}{h_n}\right)\tau^{-2}$$

$$(4.141) \qquad \frac{\varrho}{\varrho_n} = \frac{h}{h_n} = \left[\left(\frac{Z}{Z_n}\right)^2 \left(\frac{1-F}{1-F_n}\right)^\alpha \tau^{-2} \right]^{\frac{1}{\Gamma-1-\alpha}}$$

$$(4.142) \qquad \frac{T}{T_n} = \left(\frac{g}{g_n}\right)\left(\frac{h}{h_n}\right)^{-1} = \left(\frac{Z}{Z_n}\right)^2 \tau^{-2}$$

For Lagrangian time profiles one obtains, using first Eqs. (4.131) and (4.132),

$$(4.143) \qquad \frac{t}{t_n} = \tau = \left[\left(\frac{Z}{Z_n}\right)^2 \left(\frac{1-F}{1-F_n}\right)^{\Gamma-1} \varphi^{\frac{\lambda+2}{2}} \right]^{\frac{\gamma}{2(\gamma-1)}}$$

$$(4.144) \qquad \frac{r}{r_n} = x = \varphi\,\tau^{\frac{2}{\lambda+2}}$$

$$(4.145) \qquad \frac{u}{u_n} = \frac{f}{f_n} = \frac{F}{F_n}\frac{x}{\tau}$$

$$\frac{p}{p_n} = \frac{g}{g_n} = \left(\frac{h}{h_n}\right)^\gamma \tag{4.146}$$

$$\frac{\rho}{\rho_n} = \frac{h}{h_n} = \left[\left(\frac{1-F}{1-F_n}\right)x^{j+1}\right]^{-1} \tag{4.147}$$

$$\frac{T}{T_n} = \left(\frac{g}{g_n}\right)\left(\frac{h}{h_n}\right)^{-1} = \left(\frac{Z}{Z_n}\right)^2 . \tag{4.148}$$

The constant of integration appearing in all the above three sets of equations are, from the boundary conditions Eqs. (4.102) and (4.104), given by

$$F_n = f_n = 1 - \beta = \frac{2}{\gamma+1} \quad \text{and} \quad Z_n = \frac{g_n}{h_n} = (1-\beta)\beta = \frac{2(\gamma-1)}{(\gamma+1)^2} \tag{4.149}$$

For the specific case of a decaying constant energy point explosion propagating into a cold atmosphere ("zero counter-pressure") of a perfect gas, with γ = const.,

$$\lambda = j + 1 + \rho'_a = \text{const} \tag{4.150}$$

while $\omega_E = 0$. This fact can be verified directly by inspection of Eq. (4.33) where, under such circumstances, the first term vanishes, while the second is annihilated by the postulate of self-similarity. In the absence of energy sources, as implied by the constant energy condition, it follows then that

$$(4.151) \qquad h\,x^j\left[\left(\sigma + \frac{f^2}{2}\right)(x - f) - \frac{fg}{h}\right] \;=\; \text{const.}$$

corresponding to the assumption of constant energy
integral Eq. (4.117), that is usually applied in this
case.

Moreover, as a consequence of the assumption that the internal energy of the undisturbed
medium is negligible, the constant in Eq. (4.151) is
zero and this equation yields then an algebraic expression for the integral curve in the phase plane,
thus gratly simplifying the solution. In terms of
the reduced coordinates, it becomes, then, for a perfect gas with constant specific heats, where $\gamma = \Gamma$,

$$(4.152) \qquad Z^2 = \frac{\gamma - 1}{2\gamma}\, F^2\, \frac{1 - F}{F - \gamma^{-1}}$$

which satisfies, of course, the governing differential
equation, Eq. (4.128).

As a consequence of this Eq. (4.124)
is reduced to :

$$(4.153) \qquad \frac{d\ln\varphi}{d\ln F} = \frac{\nu}{\lambda + 2}\, \frac{(\gamma - 1)F^2 + 2(F - \gamma^{-1})(1 - F)}{(F - \gamma^{-1})(\nu - F)}$$

which can be integrated immediately, yielding :

$$\varphi = \left(\frac{F}{F_n}\right)^{\frac{2}{\lambda+2}}\left(\frac{F-\gamma^{-1}}{F_n-\gamma^{-1}}\right)^{c_1}\left(\frac{\gamma-F}{\gamma-F_n}\right)^{c_2} \qquad (4.154)$$

where

$$C_1 = \frac{(\gamma-1)\nu}{(\lambda+2)(\nu\gamma-1)} \quad \text{and} \quad C_2 = -\frac{2+(\gamma+1)(\nu\gamma-2)\nu}{(\lambda+2)(\nu\gamma-1)}$$

$$(4.155)$$

It should be noted that a complete
set of algebraic equations for self-similar solutions
has been obtained by von Neumann(3) in 1941. His
analysis was based on a purely Lagrangian formulation
of the blast wave equations, and it was restricted to
the case of an adiabatic, point explosion in a "cold"
atmosphere of a perfect gas with constant specific
heats and uniform density. With these restrictions,
however, it yielded, in effect, all the three sets
of profiles presented in our final example, although
the algebraic expressions were quite different than
here, since they were associated with the use of a
completely different parameter. A similarly complete
set of algebraic equations for the self-similar
Eulerian space profiles, including the power law
density variation in the atmosphere into which the
wave propagates, were presented by Sedov(6) as well

as Korobeinikov et al. (9) who derived also a simple transformation yielding the Lagrangian time profiles, but restricted in validity to the case of a perfect gas with constant specific heats. Their results were based on the use of a similar parameter, F , as here, with the only difference, however, that it was limited in scope by the condition of $\tau = 1$, that is to the form it has just for the Eulerian space profiles.

For illustration, specific solutions for the constant energy point explosion in a cold atmosphere of a perfect gas are depicted here in graphical form, representing, in effect, respectively, the plots of Eqs. (4.133)-(4.148) for the particular case of $\varphi = \varphi(F)$ specified by Eq. (4.152). Thus Figs. 4.2 - 4.5 (see pages 211 and 212) show the Eulerian space profiles, Figs. 4.6 - 4.9 (see pages 213 and 214), the Eulerian time profiles, and Figs. 4.10 - 4.13 (see pages 215 and 216) the Lagrangian time profiles, the first two in each group describing the dependence of the profiles of the various gasdynamic parameters (including the corresponding values of the parameter F) on the specific heat ratio, γ , for spherical geometry of the flow field ($j = 2$), while the remaining two depict the dependence on the geometry of the flow field in the case of a perfect gas with $\gamma = 1.4$.

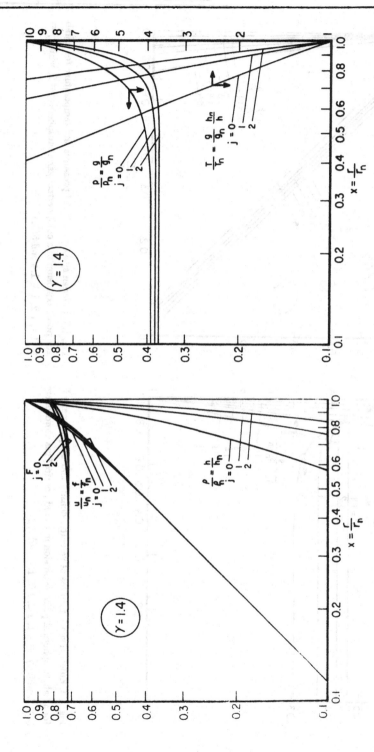

Fig. 4.2. Eulerian space profiles of velocity, density and the parameter F for adiabatic point explosion in a cold, perfect gas atmosphere (j = 0, 1 and 2, while γ = 1.4).

Fig. 4.3. Eulerian space profiles of pressure and temperature for adiabatic point explosion in a cold, perfect gas atmosphere (j = 0, 1 and 2, while γ = 1.4.)

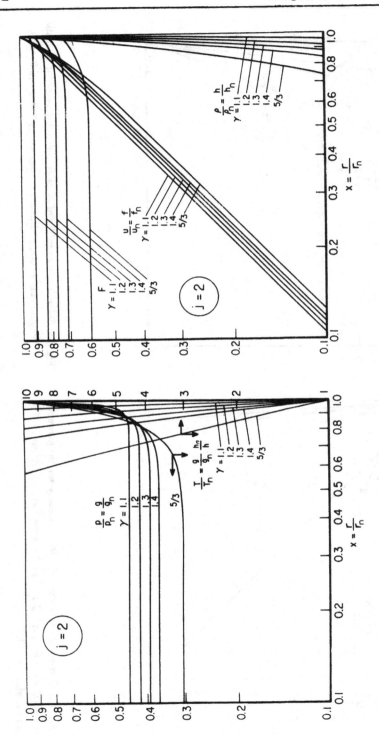

Fig. 4.5. Eulerian space profiles of pressures and temperature for adiabatic point explosion in a cold, perfect gas atmosphere ($j = 2$, while $\gamma = 1.1, 1.2, 1.3, 1.4$ and $5/3$.)

Fig. 4.4. Eulerian space profiles of velocity, density and the parameter F for adiabatic spherical point explosion in a cold, perfect gas atmosphere ($j = 2$, while $\gamma = 1.1, 1.2, 1.3, 1.4$, and $5/3$).

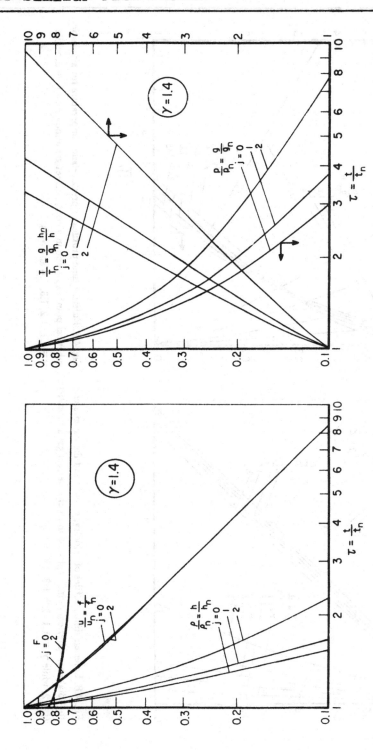

Fig. 4.7. Eulerian time profiles of pressure and temperature for adiabatic point explosion in a cold, perfect gas atmosphere ($j = 0$, 1 and 2, while $\gamma = 1.4$).

Fig. 4.6. Eulerian time profiles of velocity, density and the parameter F for adiabatic point explosion in a cold, perfect gas atmosphere ($j = 0$, 1 and 2, while $\gamma = 1.4$).

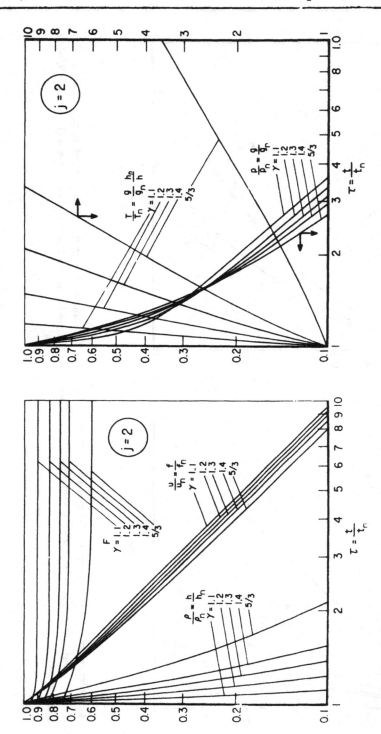

Fig. 4.9. Eulerian time profiles of pressure and temperature for adiabatic spherical point explosion in a cold, perfect gas atmosphere ($j = 2$, while $\gamma = 1.1, 1.2, 1.3, 1.4$ and $5/3$).

Fig. 4.8. Eulerian time profiles of velocity, density and the parameter F for adiabatic, spherical point explosion in a cold, perfect gas atmosphere ($j = 2$, while $\gamma = 1.1, 1.2, 1.3, 1.4$ and $5/3$).

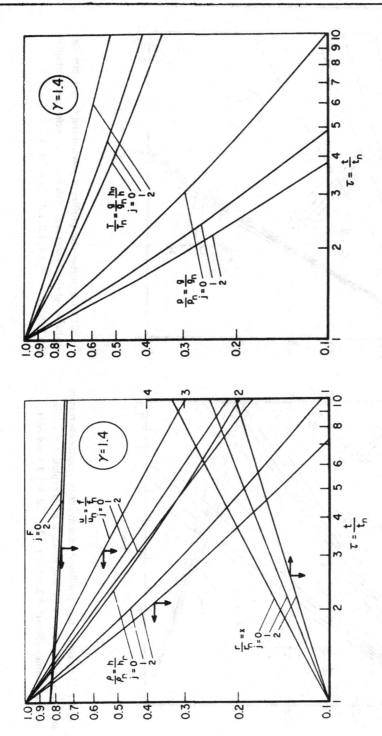

Fig. 4.11. Lagrangian time profiles of pressure and temperature for adiabatic point explosion in a cold, perfect gas atmosphere ($j = 0$, 1 and 2, while $\gamma = 1.4$).

Fig. 4.10. Lagrangian time profiles of velocity, density, space coordinate x and the parameter F for adiabatic point explosion in a cold, perfect gas atmosphere ($j = 0$, 1 and 2, while $\gamma = 1.4$).

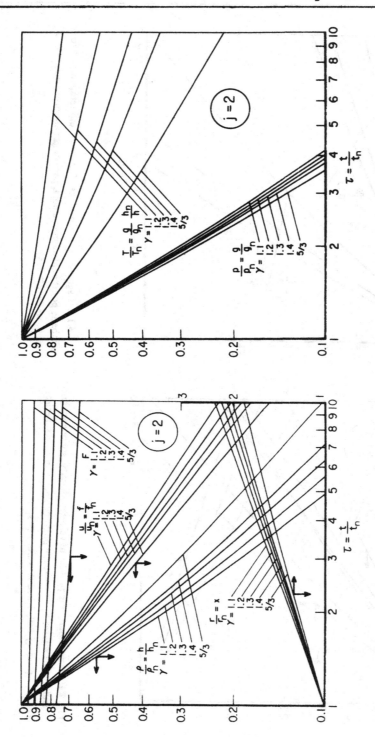

Fig. 4.13. Lagrangian time profiles of pressure and temperature for adiabatic, spherical point explosion in a cold, perfect gas atmosphere ($j = 2$, while γ = 1.1, 1.2, 1.3, 1.4 and 5/3).

Fig. 4.12. Lagrangian time profiles of velocity, density, space coordinate and the parameter F for adiabatic, spherical point explosion in a cold, perfect gas atmosphere ($j = 2$, while γ = 1.1, 1.2, 1.3, 1.4 and 5/3).

References.

1. Taylor, Sir Geoffrey, "The formation of a Blast Wave by a Very Intense Explosion", first published in British Report RC-210, June 27, 1941 ; revised version in Proc. Roy. Soc., London, A201, Part I, pp. 159-174, Part II, pp. 175-186, March 1950.

2. Taylor, G.I., "The Air Wave Surrounding an Expanding Sphere", Proc. Roy. Soc., London, A186, pp. 273-292, 1946.

3. Von Neumann, J., "The Point Source Solution", first published in NDRC, Div. B. Report AM-9, June 30, 1941 ; then in Shock Hydrodynamics and Blast Waves (ed. H. A. Bethe) AECD-2860, 1944 ; revised version in Blast Waves (es. H.A. Bethe) Los Alamos Sci. Lab. Rep. LA-2000, 27-55, 1947 ; reprinted in John von Neumann Collected Works (ed. A.H. Taub), VI, pp. 219-237, Pergamon Press, New York, 1963.

4. Von Neumann, J., and Goldstine, H., "Blast Wave Calculation", Com. Pure Appl. Math., 8, pp. 327-353, 1955 ; reprinted in John Von Neumann Collected Works (ed. A.H. Taub), VI, pp. 386-412, Pergamon Press, New York, 1963.

5. Sedov, L.I., "Rasprostraneniye sil'nykh vzryvnykh voln" (Propagation of Intense Blast Waves), Priklednaya matematika i mekhanika (Applied Mathematics and Mechanics)

$\underline{10}$, 2, 1946.

6. Sedov,L.I., "Rasprostraneniye sil'nykh vzryvnykh
       voln", revised version in $\underline{\text{Similarity}}$
       $\underline{\text{and Dimensional Method in Mechanics}}$,
       Fourth Printing, Gostekhizdat, Moscow,
       1947 (transl : M. Friedman, Ed. : M.
       Holt, Academic Press, New York,XVI +
       363 pp., 1959).

7. Stanyukovich, K.P., "Application of Particular
       Solutions for Equations of Gasdynamics
       to the Study of Blast and Shock Waves",
       $\underline{\text{Rep. Acad. Sci. USSR}}$, $\underline{52}$, 7, 1946.

8. Stanyukovich, K.P., $\underline{\text{Unsteady Motion of Continuous}}$
       $\underline{\text{Media}}$, Moscow, Gostekhizdat, 1955
       (English translation, ed. M. Holt, 1960,
       New York, Pergamon Press, XIII + 745
       pp.).

9. Korobeinikov, V.P., Mil'nikova, N.S. and Ryazanov,
       Ye. V., $\underline{\text{The Theory of Point Explosion}}$;
       Fizmatgiz, Moscow, 330 pp., 1961
       (English translation, U.S. Dept. of
       Commerce, JPRS : 14, 334, CSO : 6961-
       N, Washington, 556 pp., 1962).

10. Sakurai, A., "Blast Wave Theory", $\underline{\text{Basic Develop-}}$
       $\underline{\text{ments in Fluid Dynamics}}$, $\underline{I}$. (ed. M.
       Holt), Academic Press, New York pp.
       309-375, 1965.

11. Courant, R. and Friedrichs, K.O., $\underline{\text{Supersonic Flow}}$
       $\underline{\text{and Shock Waves}}$, John Wiley and Sons,
       New York, XVI + 464 pp., 1948.

12. Zel'dovich, Ya.B. and Raizer, Yu P., $\underline{\text{Physics of}}$
       Shock Waves and High-Temperature Hydro-
       dynamic Phenomena,2nd ed., Izdatel'stuo
       Nauka, Moscow, 686 pp., 1966 ; (English

translation, ed. W.D. Hayes and R.F. Probstein, Academic Press, New York, $\underline{I}$, XXIV + 464 pp., $\underline{II}$, XXIV + 465-916 pp., 1967).

13. Brode, H.L., "Numerical Solutions of Spherical Blast Waves", J. Appl. Physics, $\underline{26}$, 6, 766-775, 1955.

14. Brode, H.L. "Gasdynamic Motion with Radiation : A General Numerical Method", Astronautica Acta, $\underline{14}$, 5, 433-444, 1969.

15. Grigorian, S.S., "Cauchy's Problem and the Problem of a Piston for One-Dimensional Non-Steady Motions of a Gas", J. Appl. Math. Mech. , $\underline{22}$, 2, 187-197, 1958.

16. Oshima, K., "Blast Waves Produced by Exploding Wires", Exploding Wires, $\underline{II}$(ed. W. Chace and H. Moore) Plenum Press, Ind., New York, pp. 159-174, 1962.

17. Korobeinikov, V.P. and Cuskin, P.I., "Plane, Cylindrical and Spherical Blast Waves in a Gas with Counter-Pressure", Proc. V.A. Steklov Inst. of Math. (in Non-Steady Motion of Compressible Media Associated with Blast Waves", ed. L.I. Sedov) Izdatel-stvo "Nauka", Moscow, pp 4-33, 1966.

18. Levin, V.A. and Chernyi, G.G., "Asymptotic Laws of Behavior of Detonation Waves", Prikladnaya Matematika i Mekhanika, $\underline{31}$, 3, 393-405, 1967.

19. Bishimov, E., Korobeinikov, V.P., Levin V.A. and Chernyi, G.G., "One dimensional Non-Steady Flow of Combustible Gas Mixtures Taking into Account Finite Rates of

Chemical Reactions," Proc. USSR Acad. Sci. 6, 7-19, 1968.

20. Korobeinikov, V.P., "The Problem of Point Explosion in a Detonating Gas", Astronautica Acta, 14, 5, 411-420, 1969.

21. Laumbach, D.D. and Probstein, R.F., "A point Explosion in a Cold Exponential Atmosphere", J. Fluid Mech., 35, 1, 53-75, 1969.

22. Lee, J.H., "Collapsing Shock Waves in a Detonating Gas", Astro. Acta, 14, 7, 421-425, 1969.

23. Jeffrey, A. and Taniuiti, T., Non-linear Wave Propagation with Applications to Physics and Magnitohydrodynamics, Academic Press, New York, IX + 369 pp., 1964.

Printed in the United States
By Bookmasters

Printed in the United States
By Bookmasters